Proteases and Cancer

Series Editor
John M. Walker
School of Life Sciences
University of Hertfordshire
Hatfield, Hertfordshire, AL10 9AB, UK

For other titles published in this series, go to
www.springer.com/series/7651

METHODS IN MOLECULAR BIOLOGY™

Proteases and Cancer

Methods and Protocols

Edited by

Toni M. Antalis* and Thomas H. Bugge†

*Center for Vascular and Inflammatory Diseases and the Department of Physiology,
University of Maryland School of Medicine, Baltimore, MD, USA
† Oral and Pharyngeal Cancer Branch, National Institute of Dental and Craniofacial Research,
National Institutes of Health, Bethesda, MD, USA

Humana Press

Editors

Toni M. Antalis
Center for Vascular and Inflammatory
    Diseases and the Department of Physiology
University of Maryland School
    of Medicine
Baltimore, MD
USA
tantalis@som.umaryland.edu

Thomas H. Bugge
Oral and Pharyngeal Cancer Branch
National Institute of Dental and
    Craniofacial Research
National Institutes of Health
Bethesda, MD
USA
tbugge@mail.nih.gov

ISBN: 978-1-60327-002-1   e-ISBN: 978-1-60327-003-8
ISSN: 1064-3745   e-ISSN: 1940-6029
DOI: 10.1007/978-1-60327-003-8

Library of Congress Control Number: 2008924784

Preface

The recent availability of the complete genomic sequences of several mammalian organisms has led to an explosion in the knowledge of proteolytic enzymes. The realization that proteases constitute more than 2% of the total genes in the human genome underscores the importance of proteolysis to human biology and to human diseases such as cancer. Cancer development is characterized by genetic alterations in critical genes, including oncogenes and tumor suppressor genes, that provide proliferative or survival advantages to the developing tumor. Tumor cells further epigenetically subvert normal physiological processes in order to promote their own growth and survival. Proteases decisively contribute to cancer development and promotion by regulating the activities of growth factors/cytokines and signaling receptors, as well as the composition of the extracellular matrix, thereby suppressing cell death pathways and activating cell survival pathways. In addition, proteases are recognized to contribute to malignant potential in several tissue contexts by stimulating tumor cell migration and invasion. The tumor microenvironment, which is increasingly recognized to play a critical role in tumor development and progression, is critically modulated by protease activities, through activation of tumor-promoting inflammatory, angiogenic, and immune pathways. Thus, effective targeting of specific protease activities associated with tumor development holds considerable promise for improving the diagnosis and treatment of cancers.

Proteases are enzymes that catalyze the cleavage of proteins through the hydrolysis of peptide bonds. The known array of proteases and protease homologues comprises five protease families known as metalloproteinases, serine proteases, cysteine proteases, aspartic acid proteases, and threonine proteases. Protease cleavage is usually irreversible; therefore, the activities of proteases must be tightly regulated in order to prevent inappropriate and frequently pathogenic proteolysis. Most proteases are either synthesized as inactive zymogens or they are sequestered in latent forms that are specifically activated to unleash cellular responses critical for normal physiology. The sequelae of biological responses that may be generated by such protease zymogen cascades are frequently underestimated. Proteases are further controlled by regulated access of proteases to their appropriate substrates and spatial compartmentalization of proteases in cells and tissues. The regulated termination of proteolytic activities is also critical and may be accomplished by specific endogenous inhibitors and clearance receptors. Because of the complexity and intricate regulation of protease systems, delineating the activities and specific biological impact of protease pathways will require refinement of current technologies and development of unique molecular tools.

The goal of this volume is to bring together a wide range of complimentary techniques that have been developed for the specific detection and analysis of proteases and their activities in cancer. While a variety of techniques are now available for detection and imaging of proteins and receptors in many cellular contexts, the technologies for determining specific protease expression, activity, and functionally important downstream targets remain relatively poorly developed when compared to other classes of enzymes. As a prelude to understanding the approaches taken by investigators toward understanding the broad biological functions of proteases in the pathogenesis of cancer, we first present an overview of major protease classes implicated in cancer development and the issues involved in delineating protease functions during cancer development. The complexity of the roles played by secreted, pericellular, and intracellular proteases is highlighted. Examples of mechanistic studies using organ-specific mouse models of cancer development combined with transgenic targeting of proteases highlight the diverse and complex roles that proteases may play in tumor initiation and progression.

The spectrum of proteases that are expressed at a specific moment or circumstance by a cell, tissue, or organism has been termed the "degradome." New insights from genomic and computational analyses of degradomes and the challenges faced in the analysis of the large amount of information and techniques for simplifying the complexity of protease cascades are discussed next. This analysis has provided unique perspectives into the evolution of mammalian organisms. The spectrum of proteases, protease inhibitors, and protease interactors associated with cancer development can be experimentally identified at the transcript level by the profiling of gene expression using microarray technologies. The study of the expression of protease genes involved in cancer development presents additional unique challenges, however, since proteases are expressed not only by the tumor cells, but also by stromal cells that are located within the tumor microenvironment surrounding the tumor (e.g., fibroblasts, endothelial cells). Determination of which cells are expressing proteases and the molecules that regulate them is critical to understanding the functional roles of the proteases and designing effective therapeutic applications. An elegant approach to this challenge is presented using a microarray strategy that has the advantage of distinguishing the cellular origin of the genes detected through differential species specificity. The approach involves xenograft models in which human tumors are implanted into mice. Tissues are analyzed using a dual species protease/inhibitor microarray chip, validated for its ability to distinguish between human and mouse transcripts.

Proteases are associated with specific unique recognition sequences where the substrate peptide bond is cleaved, designated ...P1-P1'..., where cleavage occurs between P1-P1' amino acid residues. Biochemical and proteomics methods have proved invaluable in the identification of preferred cleavage sequences, which can then in turn be used in the design of experimental tools for detection or inhibition of protease activity and to assist in predicting endogenous substrates and signaling pathways that involve the protease of interest. Ideally, these approaches must be rapid yet comprehensive with respect to the inclusion of all potential target recognition sequences. Presented in this volume are several complementary high-throughput approaches aimed at identifying preferred cleavage sequences for proteases. In the first, positional scanning synthetic combinatorial libraries of fluorogenic peptides have been developed to identify the preferred amino acid residues at P1 through P4 positions for a given protease. This strategy has proven extremely useful for determining preferred cleavage sequences specific to a protease; however, possible interdependence between positions may not be identified by this method, since every position other than the position of interest is randomized. In an alternative strategy, protease cleavage site preferences are determined using mixture-based peptide libraries where random amino-terminally capped peptide mixtures are digested with the protease of interest and the cleavage products analyzed by automated Edman sequencing. Based on initial determinations, the process is reiterated until the full cleavage motif preference of the protease is known. In the final strategy, a substrate phage display technique is presented where randomized peptide substrates are displayed as fusion proteins on the outside of a bacteriophage. Here, the technique has been adapted to a semiautomatic platform in order to obtain a rapid but comprehensive substrate recognition profile for a given protease.

Knowledge of the preferred cleavage sequence of the given protease is the first step toward enabling detection of protease activity. Proteolysis within the pericellular environment occurs predominantly on the cell surface, where inactive zymogens, protease inhibitors, and other protease interactors assemble into multiprotein complexes that mediate the sequential and spatially restricted activation of protease zymogens and their cleavage of target substrates. Therefore, a critical aspect of our understanding of proteases in cancer biology is the ability to detect expression of specific protease activities on cells. A novel method for revealing protease activity on single living cells is presented, which employs a simple noninvasive assay for detecting the presence of specific cell surface proteolytic activity that is based on the use of modified bacterial cytotoxins.

Organ and tumor-type specific regulation of protease activities and substrates is an important consideration in our understanding of protease functions in cancer biology. Preferred cleavage sequences reveal only what substrates can be cleaved, not necessarily what substrates are cleaved in vivo. Identifying and verifying biologically relevant substrates is a major challenge for the protease field. An approach to this problem is presented where differential isotopic labeling is applied to selectively identify potential protein substrates in complex protein mixtures, followed by identification and quantification by tandem mass spectrometry proteomics and database searching.

There is an urgent need to develop more effective strategies for earlier diagnosis of cancer, earlier detection of tumor recurrence, and therapeutic targeting of tumors. Ideally, new diagnostic technologies should be directed toward noninvasive in vivo imaging, which would enable more efficient patient screening and better patient management, thus leading to improved outcomes. Several approaches for catalytic targeting of protease activities associated with cancers that have potential for both the diagnosis and treatment of tumors are presented. In the first, a noninvasive molecular imaging approach is employed for detection of tumor-associated proteolytic activity in living animals. This optical imaging approach utilizes visible and near-infared fluorescence resonance energy transfer (FRET) fluorophore pairs and has been successfully applied to detect and measure matrix metalloproteinase proteolytic activity in tumors in mouse models of cancer. Antiprotease therapy for cancer is a potentially powerful strategy to specifically target rate-limiting steps or proteolytic pathways within tumor cells. An elegant strategy involving the reengineering of protease-activated anthrax toxins to eliminate tumor cells, which has shown promise, is presented. This approach has been demonstrated to successfully target tumor cells that exploit the uPA and MMP protease pathways on the tumor cell surface is next presented, whereby the anthrax toxin PrAg-furin cleavage site is reengineered to generate a PrAg cleaved by the protease of interest. Finally, the development of protease-cleavable linkers that may enable rapid clearance from the circulation of radioisotopes "on demand" is presented, with clinical potential for radioimmunotherapeutic approaches to the treatment of cancer.

The editors are grateful to the contributing authors for providing their expertise in order to bring this volume to fruition. We are also grateful to our all of our colleagues in the protease community who have contributed thoughtful insights into the biological and biochemical approaches discussed within these pages.

Baltimore, MD   *Toni M. Antalis*
Bethesda, MD   *Thomas H. Bugge*

Contents

Contributors

NESRINE I. AFFARA • *Department of Pathology, University of California, San Francisco, San Francisco, CA, USA*

PAULINE ANDREU • *Department of Pathology, University of California, San Francisco, San Francisco, CA, USA*

TONI M. ANTALIS • *Center for Vascular and Inflammatory Diseases and the Department of Physiology, University of Maryland School of Medicine, Baltimore, MD, USA*

THOMAS H. BUGGE • *Oral and Pharyngeal Cancer Branch, National Institute of Dental and Craniofacial Research, National Institutes of Health, Bethesda, MD, USA*

PIOTR CIEPLAK • *Bioinformatics Program, Burnham Institute for Medical Research, La Jolla, CA, USA*

LISA M. COUSSENS • *Department of Pathology, Helen Diller Family Comprehensive Cancer Center University of California, San Francisco, San Francisco, CA, USA*

CHARLES S. CRAIK • *Department of Pharmaceutical Chemistry, The University of California, San Francisco, San Francisco, CA, USA*

LEONARD J. FOSTER • *Centre for High-Throughput Biology and Department of Biochemistry and Molecular Biology, University of British Columbia, Vancouver, BC, Canada*

ARTHUR E. FRANKEL • *Cancer Research Institute of Scott & White Memorial Hospital, Temple, TX, USA*

MAGDA GIOIA • *Departments of Oral Biological and Medical Sciences and Biochemistry and Molecular Biology and the Centre for Blood Research, University of British Columbia, Vancouver, BC, Canada*

JOHN P. HOBSON • *Meso Scale Diagnostics, Gaithersburg, MD, USA*

PAPPANAICKEN R. KUMARESAN • *Division of Hematology & Oncology, Department of Internal Medicine, UC Davis Cancer Center, University of California Davis, Sacramento, CA, USA*

KIT S. LAM • *Division of Hematology & Oncology, Department of Internal Medicine, UC Davis Cancer Center, University of California Davis, Sacramento, CA, USA*

STEPHEN H. LEPPLA • *Laboratory of Bacterial Diseases, National Institute of Allergy and Infectious Diseases, National Institutes of Health, Bethesda, MD, USA*

SHIHUI LIU • *Laboratory of Bacterial Diseases, National Institute of Allergy and Infectious Diseases, National Institutes of Health, Bethesda, MD, USA*

CARLOS LÓPEZ-OTÍN • *Departamento de Bioquímica y Biología Molecular, Facultad de Medicina, Instituto Universitario de Oncología, Universidad de Oviedo, Oviedo, Spain*

JUNTAO LUO • *Division of Hematology & Oncology, Department of Internal Medicine, UC Davis Cancer Center, University of California Davis, Sacramento, CA, USA*

LYNN M. MATRISIAN • *Department of Cancer Biology, Vanderbilt University, Nashville, TN, USA*

J. OLIVER MCINTYRE • *Department of Cancer Biology, Vanderbilt University, Nashville, TN, USA*

KAMIAR MOIN • *Department of Pharmacology and Barbara Ann Karmanos Cancer Institute, Wayne State University, Detroit, MI, USA*

STEFANIE R. MULLINS • *Department of Medical Oncology, Fox Chase Cancer Center, Philadelphia, PA, USA*

GONZALO R. ORDÓÑEZ • *Departamento de Bioquímica y Biología Molecular, Facultad de Medicina, Instituto Universitario de Oncología, Universidad de Oviedo, Oviedo, Spain*

CHRISTOPHER M. OVERALL • *Departments of Oral Biological and Medical Sciences and Biochemistry and Molecular Biology and the Centre for Blood Research, University of British Columbia, Vancouver, BC, Canada*

XOSE S. PUENTE • *Departamento de Bioquímica y Biología Molecular, Facultad de Medicina, Instituto Universitario de Oncología, Universidad de Oviedo, Oviedo, Spain*

VÍCTOR QUESADA • *Departamento de Bioquímica y Biología Molecular, Facultad de Medicina, Instituto Universitario de Oncología, Universidad de Oviedo, Oviedo, Spain*

BORIS RATNIKOV • *Center on Proteolytic Pathways, Burnham Institute for Medical Research, La Jolla, CA, USA*

ERIC L. SCHNEIDER • *Department of Pharmaceutical Chemistry, The University of California, San Francisco, San Francisco, CA, USA*

DONALD SCHWARTZ • *Biodiscovery, LLC, Ann Arbor, MI, USA*

BONNIE F. SLOANE • *Department of Pharmacology and Barbara Ann Karmanos Cancer Institute, Wayne State University, Detroit, MI, USA*

JEFFREY W. SMITH • *Center on Proteolytic Pathways, Burnham Institute for Medical Research, La Jolla, CA, USA*

BENJAMIN E. TURK • *Department of Pharmacology, Yale University School of Medicine, New Haven, CT, USA*

Chapter 1

Delineating Protease Functions During Cancer Development

Nesrine I. Affara, Pauline Andreu, and Lisa M. Coussens

Summary

Much progress has been made in understanding how matrix remodeling proteases, including metalloproteinases, serine proteases, and cysteine cathepsins, functionally contribute to cancer development. In addition to modulating extracellular matrix metabolism, proteases provide a significant protumor advantage to developing neoplasms through their ability to modulate bioavailability of growth and proangiogenic factors, regulation of bioactive chemokines and cytokines, and processing of cell–cell and cell–matrix adhesion molecules. Although some proteases directly regulate these events, it is now evident that some proteases indirectly contribute to cancer development by regulating posttranslational activation of latent zymogens that then directly impart regulatory information. Thus, many proteases act in a cascade-like manner and exert their functionality as part of a proteolytic pathway rather than simply functioning individually. Delineating the cascade of enzymatic activities contributing to overall proteolysis during carcinogenesis may identify rate-limiting steps or pathways that can be targeted with anti-cancer therapeutics. This chapter highlights recent insights into the complexity of roles played by pericellular and intracellular proteases by examining mechanistic studies as well as the roles of individual protease gene functions in various organ-specific mouse models of cancer development, with an emphasis on intersecting proteolytic activities that amplify programming of tissues to foster neoplastic development.

Key words: ADAMs, Angiogenesis, Cysteine cathepsins, Cancer, ECM remodeling, Inflammation, Metalloproteinases, Mouse models, Plasminogen activators, Proteases, Proteolytic cascades, Serine proteases.

1. Introduction

A unifying concept of cancer development is the acquisition of genetic alterations in critical genes, including oncogenes and tumor suppressor genes that provide a survival and/or proliferative advantage to mitotically active cells. However, it is now

Toni M. Antalis and Thomas H. Bugge (eds.), *Methods in Molecular Biology, Proteases and Cancer, Vol. 539*
© Humana Press, a part of Springer Science+Business Media, LLC 2009
DOI: 10.1007/978-1-60327-003-8_1

clear that genetically-altered neoplastic cells co-opt important physiologic host-response processes early during cancer development to favor their own survival, including extracellular matrix (ECM) remodeling, angiogenesis, and activation/recruitment of innate and adaptive leukocytes (inflammation). In particular, cancer development requires reciprocal interactions between genetically altered neoplastic cells and activated stromal cells, as well as the dynamic microenvironment in which they both live *(1, 2)*. Of note, proteases derived from activated host stromal cells have recently been identified as critical cofactors for cancer development. Although it was initially believed that matrix remodeling proteases merely regulated migration and/or invasion of neoplastic cells into ectopic tissues, there is growing evidence that proteases contribute to cancer development by regulating bioactivity of a myriad of growth factors, chemokines, soluble and insoluble matrix molecules that regulate activation and/or maintenance of overall tissue homeostasis, as well as inflammatory and angiogenic programs (**Fig. 1**) *(3)*.

Recent advances in activity-based profiling of protease function *(4–8)* have enabled tracking distribution and magnitude

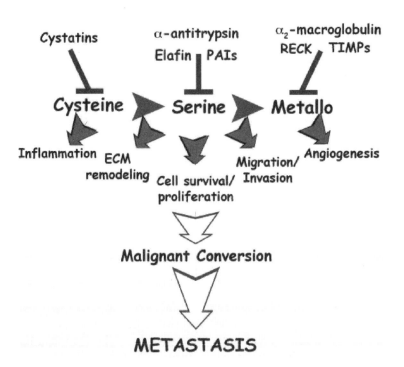

Fig. 1. Proteases act as critical cofactors for cancer development. Proteolysis, a central cofactor for neoplastic progression, results from a cascade-like activation of linear protease circuits, including key upstream proteases, such as cysteine and serine proteases, which converge leading to amplification of metalloproteinase proteolytic activities. Tumor net proteolysis contributes to tumor progression by mediating tissue remodeling, inflammation, angiogenesis, and acquisition of invasive capabilities, cell survival, and proliferation.

of proteolytic activities in cells and tissues *(9)*. Together with observations gained from examining individual protease gene functions in mouse models of de novo carcinogenesis, have emerged insights into the multitude of enzymatic activities that participate in tissue remodeling associated with cancer development *(10)*. Recently, sequencing of the human genome enabled characterization of the human degradome found to consist of at least 569 proteases and homologues that belong to various classes, including metalloproteinases (MMP), serine proteases, and cysteine cathepsins *(10)*. For many of these enzymes, their most significant protumor activity may lie in their ability to post-translationally regulate downstream proteases initially secreted as either inactive zymogens or sequestered by matrix in latent forms *(10)*. This realization has led to the notion that embedded within tissues are complex, interconnecting protease networks that, depending on the tissue perturbation, selectively engage specific protease amplification circuits *(11)*. These coordinated efforts regulate overall tissue homeostasis, response to acute damage and subsequent tissue repair, as well as contribute to pathogenesis of chronic disease states such as cancer.

2. Proteases Implicated in Cancer Development

Requisite for neoplastic cell, vascular, or inflammatory cell invasion during tumorigenic processes are the remodeling events that are initiated within tumor stroma in pericellular microenvironments. In epithelial tumors, a majority of ECM-remodeling proteases emanate from activated stromal cells, a large percentage of which being infiltrating leukocytes such as mast cells, other myeloid-lineage cells and lymphocytes *(3)*. In vivo assessment of individual protease gene functions have indeed identified some proteases as significant cofactors for cancer development because of their ability to directly regulate important aspects of neoplastic progression (**Fig. 1**), while others are significant as they set in motion interconnecting protease cascades, resulting in amplification of enzymatic activity of "terminal" proteases (**Fig. 2**). Although these cascades of proteolytic activation are crucial for tumorigenesis and resemble those regulating coagulation *(12)* and/or complement *(13)*, *in vivo* experimental studies using mouse models have revealed organ and tumor type-specific regulation of protease bioactivities, as well as involvement of proteases emanating from multiple enzymatic classes, i.e., cysteine, serine, and metallo. In the sections that follow, we discuss the diversity of proteases whose bioactivity result in amplification of *terminal* proteases in neoplastic tissues.

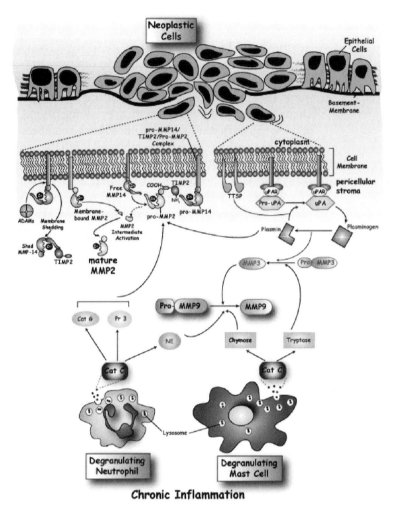

Fig. 2. Intersecting protease pathways during neoplastic progression. Proteolysis of extracellular matrix (ECM) components during tumor progression results from the activity of combined protease pathways emanating from the tumor cell compartment, including urokinase plasminogen activator (uPA), MMP14, and type II transmembrane serine proteases (TTSP), as well as proteases expressed by supporting tumor stromal cells, such as neutrophils and mast cell-derived proteases, such as MMP9, plasmin, mast cell chymase, mast cell tryptase, neutrophil elastase, and cathepsin C (Cat C). Rather than functioning individually, each protease functions as a "signaling molecule," exerting its effects as part of a proteolytic pathway, where proteases potentially interact and activate other proteases in a cascade-like manner, culminating in amplification of enzymatic activity of "terminal" proteases, such as MMP9.

2.1. Metalloprotease Activation and Function During Cancer Development

MMPs, also known as matrixins, are a family of zinc-dependent endopeptidases that act as extrinsic factors regulating critical parameters of neoplastic progression. MMPs facilitate cancer development by triggering release of growth and angiogenic factors sequestered in neoplastic tissues, as well as activation of inflammatory mediators and processing of cell–cell and cell–matrix adhesion molecules (14). To date, 23 vertebrate MMPs have been identified and classified into distinct categories based on domain structure and substrate specificity *(14, 15)*. Bioactivity of MMP function is controlled posttranslationally. Secreted MMPs (with the exception of stromelysin-3/MMP11) remain as inactive zymogens (or pro-MMPs) requiring enzymatic and/or autolytic removal of propeptide domains, rendering the active site available to cleave substrates. However, once activated, MMPs are further regulated by three major types of endogenous inhibitors, α_2-macroglobulin, RECK *(16)*, and tissue inhibitors of metalloproteinases (TIMPs) (reviewed in *(14)*). Bioactivity of TIMPs is further regulated posttranslationally where some are inactivated by serine protease cleavage *(17)*.

The most compelling evidence for MMPs as active contributors to neoplastic progression comes from tumor-prone organ-specific mouse models harboring homozygous null gene deletions or tissue-specific overexpression of individual MMPs. In a transgenic mouse model of multistage skin carcinogenesis where the early region genes of human papillomavirus type 16 (HPV16) are expressed as transgenes under control of the human keratin 14 (K14) promoter, e.g., K14-HPV16 mice *(18)*, genetic elimination of MMP9 significantly reduced the incidence of carcinomas while on the other hand, reconstitution of K14-HPV16/MMP-9[null] mice with wild type bone marrow-derived cells restored characteristics of neoplastic development and tumor incidence to levels similar to control HPV16 mice *(19)*. Likewise, angiogenesis and tumor development were significantly inhibited during pancreatic islet carcinogenesis *(20)* and cervical carcinogenesis *(21)* following genetic deletion or pharmacological inhibition of MMP9. Similarly, during development of ovarian carcinomas using a xenograph model *(22)*, as well during neuroblastoma development *(23)*, reconstitution of MMP9-deficient mice with MMP9-proficient bone marrow-derived cells restored cellular programs necessary for development of angiogenic vasculature, tissue remodeling, and overt tumor development. Thus, in each of these distinct tissue microenvironments, infiltrating leukocytes were the predominant sources of MMP9 *(19–21)*, hence, implicating leukocyte-derived MMP9 as a significant cofactor for cancer development.

In addition to MMP9, other MMPs have also emerged as important cofactors in cancer development. Studies using genetically modified mouse models have further revealed organ and tumor type-specific regulation of MMP bioactivities. For instance, overexpression of human MMP1 (collagenase) in skin suprabasal layers not only induced epidermal hyperplasia and hyperkeratosis, but also increased susceptibility to chemical skin carcinogenesis *(24)*. But then, chemically-induced skin carcinogenesis was attenuated in MMP11 null homozygous mice *(25)*, while genetic elimination of MMP7 attenuated development of intestinal adenomas in the *m*ultiple *i*ntestinal *n*eoplasia *(Min)* mouse model of intestinal neoplasia *(26)*, and overexpression of either MMP2 or MMP7 in mammary epithelium accelerated mammary tumor formation *(27, 28)*. Although these studies indicate that overexpression of a single MMP may contribute to neoplastic progression *(24, 25, 28)*, MMP3 (stromelysin 1) has been shown to contribute to spontaneous mammary neoplasms by acting as a tumor promoter in the absence of carcinogens or preexisting mutations following targeting to mammary glands *(27)*. Furthermore, although mice deficient in the transmembrane-spanning MMP (MMP14/MT1-MMP) have not been specifically examined using de novo models of cancer development, the role of MMP14 as a cancer cofactor has partially been revealed using mammary-targeted MMP14 transgenic mice that developed spontaneous mammary lesions *(29)*. Nonetheless, besides promoting tumor progression, MMPs can also exhibit anti-tumor functions. For example, using MMP3-deficient mice revealed a protective role for MMP3 during chemically induced squamous cell carcinoma development *(30)*. Similarly, loss of MMP8 (collagenase 2) enhanced rather than reduced skin tumor susceptibility in MMP8-deficient male mice *(31)*.

MMP net pericellular proteolytic activity is also dependent on the balance between levels of secreted MMPs and their endogenous inhibitors, tissue inhibitors of metalloproteinases (TIMPs). Hence, elevated TIMP levels would be expected to inhibit cancer progression. Nonetheless, the contribution of TIMPs to tumorigenesis has been controversial. In contrast to overexpression of TIMP-1 that inhibited development of SV40 large T antigen-mediated liver carcinogenesis *(32)* as well as chemically-induced mammary tumorigenesis *(33)*, overexpression of TIMP-1 enhanced rather than suppressed skin carcinogenesis by promoting keratinocyte hyperproliferation and acquisition of chromosome instability, thus enhanced premalignant cells susceptibility to undergo malignant conversion *(34)*. Interestingly, once skin tumors develop, TIMP-1 acts by stabilizing tumor stroma without limiting malignant conversion and development of metastases *(34)*. Furthermore, recent studies implicated TIMP-1 in inducing a prometastatic microenvironment that promotes tumor cell metastasis selectively to the liver by triggering hepatocyte growth factor (HGF/scatter factor) signaling *(35)*. Moreover, TIMP-2 favors tumor development by acting as an adaptor molecule that induces the formation of a plasma membrane-associated ternary complex with pro-MMP2 and MMP14 *(36–38)*, thereby promoting pro-MMP2 activation by MMP14 at the cell surface and favoring tumor progression. Several advantages to having degradative enzymes in a bound state at the cell surface have been proposed. Namely, bound proenzymes may be more readily activated, and the bound enzymes generated may be more active than the same enzymes found in the soluble phase. Bound enzymes may also be protected from inactivation by inhibitors, in addition to providing a mean of concentrating components of a multistep pathway, thereby increasing rate of reactions. Immobilizing enzymes on the cell surface or in matrix may further restrict the activity of an enzyme so that substrates only in the vicinity of the cell or adjacent matrix components are degraded. Hence, activation at the cell surface may actually provide the most significant control point in MMP activity, linking MMP expression with proteolysis and invasion.

Several mechanistic studies have supported the functional contribution of MMPs to neoplastic progression. MMPs alter the stromal microenvironment by mediating liberation of ECM sequestered growth-promoting factors, such as basic fibroblast growth factor (FGF-2), or proteolytic cleavage of growth factor latent precursors, including members of the epidermal growth factor (EGF) family such as transforming growth factor-α (TGF-α), which act as potent mitogens for neoplastic cells (reviewed in (39)).

MMPs have also been found to act as important positive regulators of tumor angiogenesis. Infiltration of MMP9-expressing inflammatory cells coincided with development of angiogenic vasculature in premalignant skin of K14-HPV16 transgenic mice (19). Although vascular endothelial growth factor (VEGF) was constitutively expressed in normal β-cells and at all stages of islet carcinogenesis, it only became bioavailable for interaction with its receptor on microvascular endothelial cells following infiltration of leukocytes expressing MMP9, thereby triggering activation of angiogenic programs (20). During development of experimental neuroblastomas, MMP9 not only regulates bioavailability of VEGF, but also mediates pericyte recruitment to developing angiogenic vessels, thus inducing stabilization of newly formed tumor vasculature (40, 41). An additional line of evidence supporting a role for MMP9 in promoting neovascularization comes from studies reporting the unique ability of MMP9 to induce release of soluble kit-ligand, which thereby initiated mobilization of hematopoietic stem cells/progenitor cells in bone marrow (42, 185).

Furthermore, MMP9 induces a tissue microenvironment that is permissive for primary tumor development, and its bioactivities also regulate secondary metastasis formation. Studies by Matrisian and colleagues have determined that MMP9 derived from inflammatory cells (possibly neutrophils) present in premetastatic lung facilitates survival/establishment of early metastatic cells, but not growth of metastatic foci (43), whereas MMP9 derived from Kupfer cells in liver parenchyma facilitates early establishment and growth of colorectal metastases to liver. Additional clues into the later events regulating metastasis have implicated MMP9 secreted by macrophages and alveolar VEGF receptor (VEGFR1)[+]-endothelial cells in microenvironmental remodeling that is crucial for metastatic cell survival in the lung *(44)*. Using a mouse model of experimental metastasis formation, Hiratsuka et al. reported that following recruitment to sites of primary tumor growth, macrophages circulate to distal organs. On the one hand, distal organs exhibiting low-level expression of VEGFR1 fail to induce MMP9 in response to leukocyte presence and are therefore not suitable environments for subsequent metastatic cell growth. On the other hand, distal organs that are VEGFR1-positive and contain a population of endothelial cells capable of inducing expression of MMP9 above that supplied by circulating macrophages are "fertile" sites for productive metastatic growth. Although induced expression of the VEGFR1 ligand VEGF-A does not appear to be involved, presence of an active VEGFR1 tyrosine kinase domain is necessary; thus, it seems reasonable that activated MMP9 releases matrix-sequestered VEGF-A rendering it bioavailable for interaction with its receptors as has been reported by Bergers and colleagues *(20)*, thus stimulating efficient vascular remodeling and angiogenesis necessary for metastatic cell growth and survival. Taken together these findings indicate that mechanisms by which premetastatic niches enhance metastatic outgrowth are organ and cancer type specific.

MMPs have also been found to enhance tumor cell migration by altering cell–cell and cell–matrix interactions. For instance, MMP14 and MMP2 have been shown to release cryptic fragments of laminin-5 γ2 chain domain III, which, due to presence of EGF-like repeats, binds to EGF-receptor on tumor cells, thus activating downstream signaling events that lead to tumor cell motility *(45, 46)*. The cell–cell adhesion junction E-cadherin has also been found to act as a substrate for MMP3 in mammary epithelial cells, triggering progressive phenotypic conversion of normal epithelial cells into invasive mesenchymal phenotype, characterized by dissolution of stable cell–cell contacts, down-regulation of cytokeratins, and induction of vimentin expression *(47)*.

It is not surprising that MMPs have attracted significant attention as anti-cancer therapeutic targets. Unfortunately, clinical evaluation of MMP inhibitors revealed no efficacy in patients suffering from the advanced stages of various types of cancer *(48)*. Nonetheless, these failed clinical trials have enabled revisiting of the upstream regulatory mechanisms controlling activation of important proteases, like MMP9, that play a clear and undisputed role in cancer development. Active MMP9 represents a terminal protease whose proteolytic activity is amplified by several proteolytic pathways that converge or act in parallel to activate the latent pro-form of the enzyme (**Fig. 2**). Indeed, serine proteinases, such as plasmin or urokinase-type plasminogen activator (uPA), neutrophil elastase, mast cell chymase, and trypsin, cleave propeptide domains of secreted pro-MMPs, such as MMP9, and consequently induce autocatalytic activation of MMPs *(14)*. Some activated MMPs can further activate other pro-MMPs. For example, MMP3 activates pro-MMP1 and pro-MMP9, whereas pro-MMP2 is resistant *(14)*. Thus, some serine proteinases act as initiators of activation cascades regulating bioactivity of pro-MMPs *in vivo*. Collectively, these observations indicate that anti protease-based therapeutics may achieve better efficacy when targeting a "pathway" as opposed to a single class or single species of enzyme(s).

2.2. ADAMs: Emerging Roles in Tumorigenesis

ADAMs (A Disintegrin And Metalloproteinase domains) are a family of cell surface proteins related to snake venom metalloproteases (SVMPs) and MMPs, characterized by the presence of both a disintegrin and metalloproteinase domains responsible for their wide range of biological activities, including proteolysis, adhesion, and signaling *(49)*. More than 30 ADAM orthologues have been identified in various species, in particular 29 are found in mammals, of which several members are expressed exclusively in the testis where they participate in spermatogenesis and fertilization. ADAMs are typically composed of the following structural and functional conserved domains: a prodomain acting as an intramolecular chaperone responsible for the protease domain inhibition, a metalloprotease domain, a disintegrin domain shown to interact with integrins to regulate cell–cell adhesion, a cysteine-rich region implicated in both cell–cell and cell–ECM interactions because of its abilities to link heparan sulfate proteoglycan, an EGF repeat domain, a transmembrane, and a cytoplasmic domain *(49–51)*. Despite the presence of a metalloprotease domain, only 17 of the 29 mammalian ADAMs identified at this date contain the catalytic consensus motif for metalloproteinases, indicating that half of the ADAMs have actual catalytic activities that have been formally proven to exist for ADAMs 10, 12, 17, and 28 *(51–55)*.

ADAMs expression has been found to be induced in human cancers, particularly ADAMs 9, 10, 12, 15, and 17, which are expressed at low levels in normal tissue but overexpressed in a variety of tumors, including breast, gastric, colon, prostate and pancreas carcinomas, non-small cell lung cancer, liver metastases, glioblastomas, as well as hematological malignancies *(56–63)*. Moreover, studies have demonstrated a predictive role of ADAMs expression for human breast cancer, as ADAM17 expression levels predict poor prognosis *(64)* and urinary concentration of ADAM12 correlates with cancer progression *(60)*.

Unlike other protease families, a role for ADAMs in tumorigenesis has only recently begun to be explored using a growing number of de novo carcinogenesis mouse models harboring either gene deletion or tissue-specific overexpression of individual ADAMs. Using the W[10] mouse model of prostate cancer where

SV40 large T antigen is expressed under the control of the probasin promoter, homozygous deletion of ADAM9 inhibited cancer progression past the well-differentiated stage *(59)*. Conversely, transgenic overexpression of ADAM9 in mouse prostate epithelium was sufficient to induce epithelial hyperplasia leading to the development of neoplastic lesions after 1 year *(59)*, thus revealing a functional role for epithelial-derived ADAM9 in prostate cancer initiation and progression likely related to the ability of ADAM9 to shed FGF receptor (FGFR2iiib) and EGF both implicated in human prostate cancer development *(65, 66)*. In the same W^{10} mouse model, ADAM12 expression was restricted to a subpopulation of stromal cells where it exerts an essential role in tumor progression and development as genetic deletion of ADAM12 resulted in smaller and more differentiated neoplastic lesions *(67)*. Accordingly, driven by the mammary epithelium promoter MMTV, overexpression of either the secreted form of ADAM12 or a membrane-anchored form lacking the intracellular domain increased both tumor burden and malignancy degree of breast carcinomas in MMTV-PyMT mice *(68)*. These observations indicate that ADAM12 extracellular region, including the protease and adhesion domains, mediates ADAM12-tumor promoting capabilities, as opposed to initiation of a signaling cascade through ADAM12 intracellular tail.

In vitro, ADAM17 expression regulates glioma tumor cell invasiveness *(62)* and human pancreatic ductal adenocarcinoma proliferation and invasive abilities *(69)*. These effects may be related to cell migration, adhesion, or matrix remodeling activities of ADAMs. Several ADAMs have indeed been found to degrade specific components of ECM, i.e., ADAM12 mediates processing of gelatin, type IV collagen, and fibronectin *(60)*, ADAMs 10 and 15 induce type IV collagen and gelatin degradation *(70, 71)*, and ADAM13 has been reported to cleave fibronectin *(72)*. Thus ADAMs-mediated cleavage of ECM proteins not only fosters cancer cell migration but also induces release of ECM-sequestered growth and angiogenic factors.

Although direct evidence is still lacking, ADAMs are also suspected to exert a functional role in tumor angiogenesis. In a mouse model of retinopathy of prematurity, mice harboring a homozygous deletion of ADAM15 present an inhibition of angiogenesis when compared with the control mice, implicating ADAM15 in the control of pathogenic angiogenesis. However, although growth of implanted tumors has also been inhibited in ADAM15$^{-/-}$ mice, there was no difference in tumor vascularity *(73)*. Moreover ADAM15 and 17 are expressed in endothelial cells (EC) *(74)* where they may exert an important functional role, since using an ADAM-specific inhibitor (GLI12971) decreased EC migration, adhesion, and proliferation *in vitro* *(75)*.

One mechanism by which active ADAMs functionally contribute to neoplastic progression is through their sheddase activity, defined as the proteolytic release of ectodomains of membrane-anchored cell-surface proteins. The sheddase activity of ADAMs regulates numerous signaling proteins, including growth factors and receptors, whose activation and/or bioavailability are dependent on of several proteolytic steps. Importantly in cancer, ADAMs have been implicated in the shedding of EGFR ligands, including TGFα, EGF, HB-EGF, betacellulin, epiregulin, and amphiregulin (reviewed in *(49)*), implicating them as a major regulator of EGFR signaling. The signals initiated by EGFR-like receptors have been extensively studied since the early 1980s and demonstrated to participate in the control of differentiation, proliferation, and cell survival as well as in the development of tumors from epithelial origin *(76)*. In particular, expression of ADAM17, initially termed tumor necrosis factor-α converting enzyme (TACE), is specifically induced in breast and non-small cell lung carcinomas (NSCL) where it seems necessary for EGFR signaling *(63, 77)*. In 3D culture model of human breast cancer progression, inhibition of TACE/ADAM17 by siRNA or small molecule inhibitors reduces breast cancer cell proliferation and reverts their malignant phenotype by inhibiting EGFR signaling *(64)*. Similarly, use of a selective ADAM inhibitor in NSCL cell lines contributes to the inhibition of EGFR signaling *(63)*. Several anti-EGFR agents have been approved to treat various human cancers, but despite their efficiency, the majority of patients do not experience long-term benefit from these therapies *(78–80)*, demonstrating a need for alternative strategies to target EGFR signaling pathway. As such, ADAMs inhibitors may offer a new opportunity for pharmacological interventions by targeting upstream the receptor that could be then used in combination with others drugs.

Furthermore, because of their ability to regulate levels of chemokines and cytokines, including tumor necrosis factor-α (TNF-α), TRANCE, CX3CL-1, CXCL-16 *(81)*, ADAMs have also been implicated in the recruitment of immune cells implicated in inflammation. For instance, TNF-α is synthesized as a transmembrane precursor that is processed through the proteolytic activity of ADAM17/TACE to a soluble form. Interestingly, mice lacking TIMP-3 develop inflamed livers associated with an increase in TNF-α activity *(82)*. The mechanism includes ability of TIMP-3 to inhibit ADAM17; thus, TIMP-3 regulates ectodomain shedding of TNF-α under physiological conditions, which if perturbed, results in high levels of soluble TNF-α and development of spontaneous inflammation.

Another intriguing role for ADAMs in tumorigenesis was proposed recently, namely shedding of membrane-anchored

MMPs, including MMP14 *(83)*. As such, shed MMP14 may compete with membrane-anchored MMP14 for TIMPs, thus altering the balance of active MMPs and their inhibitors at the cell surface as well as the vicinity of tumor cells (**Fig. 2**) *(84)*. Collectively, recent observations presented here implicate ADAMs as cofactors for cancer development. These observations shed light on the importance of understanding how a tissue responds under homeostatic circumstances by differentially activating specific protease pathways, when compared with that response following pathologic challenge. Thus, although it is evident that several protease pathways may converge, the above-mentioned studies indicate existence of protease pathways that may act in parallel to foster neoplastic progression.

2.3. Serine Protease Regulation of MMP Activity During Cancer Development

Several serine proteases have been implicated as important regulators of cancer development, some of which are known regulators of MMP9 bioactivity. This protease family includes enzymes involved in mediating activation of plasminogen (urokinase-type and tissue-type plasminogen activators, uPA and tPA, respectively) *(85)*, as well as serine proteases stored in secretory lysosomes of various leukocytes, namely mast cell chymase *(86)*, mast cell tryptase *(86)*, and neutrophil elastase (NE) *(87)*. Although most secreted serine proteases emanate from host stromal cells, recent studies implicate a superfamily of cell-surface associated serine proteases, also known as Type II Transmembrane Serine Proteases (TTSP), such as matriptase/MT-SP1 and hepsin, originating exclusively from tumor cells as important regulators of cancer development *(88)*.

Serine proteases share a common characteristic that being their synthesis as inactive zymogens whose activation involves a two-step mechanism *(89, 90)*. Following synthesis and signal peptide removal during passage through the endoplasmic reticulum, a pro-protease containing an "activation dipeptide" is generated, characterized by the presence of two amino-terminal residues that blocks substrate access to the catalytic site cleft, thereby maintaining serine proteases in their latent state and preventing premature protease activation. Furthermore, serine protease proteolytic activity is tightly regulated by acidic pH, an environment that is typical of secretory granules.

2.3.1. Plasminogen Activators

Several mechanistic studies have implicated the uPA/plasmin proteolytic cascade in functionally contributing to neoplastic progression, including acquisition of a migratory and invasive phenotype by tumor cells, as well as remodeling of ECM components via activation of a number of MMPs, such as MMP9 *(91)*. Enzymatic activity of plasmin is tightly regulated by two plasminogen activators, uPA and tPA. uPA plays a crucial role in tissue remodeling, while tPA is important in vascular fibrinolysis.

Initiation of plasmin activation occurs following binding of uPA to its receptor uPAR, a glycosylphosphatidylinositol (GPI)-anchored cell membrane protein found to localize to discrete focal contacts *(92)*. Subsequently, binding of uPA to uPAR catalyzes conversion of plasminogen to its active form, plasmin. On the other hand, uPAR-bound pro-uPA is also activated by plasmin, in turn results in a feedback pathway that accelerates plasminogen activation *(93)*. Maintenance of this protease cascade depends upon the balance between uPAR-bound uPA proteolytic activity and endogenous inhibitors, plasminogen activator inhibitor-1 (PAI-1) *(94)*. PAI-1 not only regulates proteolytic activity of uPA, but also induces rapid internalization of the uPA/PAI-1/uPAR complex *(95)*, thus regulating levels of cell surface-bound uPA as well.

Despite extensive experimental data linking uPA/uPAR system to neoplastic progression, only few studies have evaluated the role of the uPA proteolytic pathway in tumor-prone organ-specific mouse models *(96, 97)*. Although the number of intestinal adenomas in the *Min* mouse model of intestinal neoplasia (Apc$^{Min/+}$) was significantly reduced following genetic elimination of uPA, proliferation and angiogenesis of established neoplastic lesions were not altered in the absence of uPA, most likely due to mechanisms involving up-regulation of cycloxygenase-2 (Cox-2) expression and the Akt signaling pathway *(96)*. These observations implicate a tumor promoting role of uPA during the early stages of intestinal adenoma development through mechanisms involving leukocyte infiltration *(96)*.

Further mechanistic studies implicated uPA/plasmin proteolytic cascade as a key event conferring tumor cells with ability to migrate through fibrinous matrices. By directly activating pro-MMPs, including pro-MMP-1, -2, -3, and -9 *(91, 98–100)*, plasmin contributes to localized extracellular proteolysis and ECM remodeling at the leading edge of migrating tumor cells. These data support the presence of a cascade of proteolytic activations that converge leading to activation of common terminal proteases, including MMP9 (**Fig. 2**).

2.3.2. Mast Cell Serine Proteases

The functional significance of mast cell-derived serine protease activity (chymases and tryptases) in neoplastic progression has recently been appreciated, given their ability to trigger a proinflammatory response as well as to induce a cascade of protease activation, culminating in activation of MMP9 (**Fig. 2**). While mast cell-derived chymases and tryptases are stored in secretory granules, their release into the extracellular milieu is triggered following activation/degranulation in response to cross-linking of FcεRI, a high affinity receptor for immunoglobulin E (IgE) *(101)*. Mast cell activation can also occur independently of IgE via mechanisms involving the complement system (C3a and C5a) *(102)*, neuropeptides such as substance P *(103)*, cytokines including stem cell factor *(104)*, as well as engagement of the Toll-like receptors *(105)*.

Although one human chymase gene belonging to the α chymase family has been identified to date, rodents express several β chymases (murine mast cell protease/mMCP-1, -2, and -4) in addition to α chymase (mMCP-5) *(106)*. Notably, chymotrypsin-like activities in murine peritoneal cells and cutaneous tissue (skin) were absent following genetic elimination of mMCP-4 *(107)*, indicating that mMCP-4 is the major source of stored chymotrypsin-like activity in these tissues. On the other hand, mast cell tryptases in rodents include mMCP-6 and mMCP-7. Following degranulation, mMCP-6 is secreted locally into the vicinity of mast cells, while mMCP-7 is released into the circulation *(108)*. A novel mast cell tryptase, denoted mMCP-11/mastin, has recently been identified in dogs, pigs, and mice *(109)*. In humans, tryptases include membrane-anchored tryptases (γ-tryptase) and soluble tryptases (α-, β-, and δ-tryptases), the latter that have also been termed mMCP-7-like tryptases because of similarities with mMCP-7 genetic organization. However, β-tryptases represent the most important tryptases following secretion, and α- and δ-tryptases are both resistant to activation because of the presence of propeptide mutations and catalytic domain defects (reviewed in *(110)*). Tryptases are known to cleave substrates at the carboxyl-terminal side of basic amino acids, while chymases exert chymotrypsin-like activity, cleaving peptides at the carboxyl-terminal side of aromatic amino acids *(111, 112)*.

Although specific gene knockouts of murine chymases and/or tryptases have yet to be evaluated in de novo mouse tumor models, their role in cancer appears clear since mechanisms delineating their functional contribution to neoplastic progression have recently been examined. For instance, injection of human chymase into skin of guinea pigs induced a significant increase in neutrophil infiltration and vascular permeability *(113)*, indicating the ability of mast cell chymase to exert proinflammatory effects. Furthermore, although pro-MMP9 represents a common terminal substrate that is activated by many upstream proteases, mast cell-derived mMCP-4 plays a crucial role in MMP9 activation, given that only the proform of MMP9 was detected in tissue extracts in the absence of mMCP-4 *(107)*. α-chymase has also been shown to indirectly mediate MMP9 activation by catalyzing cleavage and hence inactivation of free TIMP-1, as well as TIMP-1 when bound in a complex with pro-MMP9 *(17)*. Remarkably, through its ability to activate MMP9, chymase exerts indirect proangiogenic activities via regulating release of sequestered VEGF from the matrix *(86)*. Moreover, a major role of mast cell-derived chymase in regulating turnover of connective tissue components, including thrombin, fibronectin, and collagen has been revealed by evaluating skin, heart, and lung tissues in mMCP-4 homozygous null mice *(107)*. In addition to activating MMP9, studies using mMCP-4-deficient mice indicate that mMCP-4 is important but not essential for activation of pro-MMP2 *in vivo*, since reduced levels of active MMP2 have been observed in the absence of mMCP-4 when compared with wild type mice *(107)*.

Mast cell tryptases have been implicated in cancer development because of their ability to activate MMP9 indirectly by initially inducing pro-MMP3 activation *(114)*. Furthermore, mast cell-derived tryptases modulate neoplastic microenvironments during skin carcinogenesis by acting as direct mitogens for stromal fibroblasts *(115, 116)* and epithelial cells *(117)*, in addition to stimulating synthesis of α_1 pro-collagen mRNA in dermal fibroblasts *(86)*. Tryptases also act as potent proinflammatory factors, given that injection of mMCP-6, but not mMCP-7, induced neutrophil infiltration into peritoneal cavities *(118)*. Furthermore, recent studies suggest that tryptases indirectly induce leukocyte recruitment by stimulating chemokine release, such as IL-8, from endothelial and epithelial cells *(117, 119)*.

The above-mentioned findings reveal that mast cell-derived chymases and tryptases are both functionally capable of modulating the tumor microenvironment by mediating dissolution of ECM components and triggering inflammation. Significantly, chymases and tryptases are stored in mast cell secretory granules in their mature and enzymatically active forms, ready for exocytic release following mast cell degranulation, as opposed to most MMPs that are secreted as zymogens requiring pericellular activation. A cysteine protease *(120)*, cathepsin C (also known as dipeptidyl peptidase I (DPPI)), particularly abundant in mast cell secretory granules, acts as a direct upstream activator of mast cell serine proteases. These observations indicate that mast cell-derived chymases and tryptases functionally connect members of the cysteine protease family with terminal proteases, such as MMP9, indicating the presence of a cascade of proteolytic activations that may serve as potential anti-cancer intervention strategies.

2.3.3. Neutrophil-Derived Serine Proteases

Neutrophil elastase (NE), a serine protease abundantly present in neutrophil azurophilic granules, is transcriptionally activated early during myeloid development *(121, 122)*. Little is known about the role of NE in cancer progression, while the role of NE in pulmonary disorders, including emphysema and fibrosis, has been well documented *(123)*. Interest in NE during neoplastic processes stems from recent clinical reports that correlate elevated NE expression with poor survival rates in patients with primary breast cancer *(124)* and non-small cell lung cancer *(125)* as well as having recently been found to initiate development of acute promyelocytic leukemia (APL) by mediating direct catalytic cleavage of PML-RARα, a protein generated by a chromosomal translocation fusing promyelocytic leukemia (*PML*) and retinoic acid receptor-α (*RAR*α) genes *(126)*. Further mechanistic studies revealed the ability of NE to directly modulate ECM components or indirectly through initiation of protease cascades, culminating in activation of terminal MMPs, such as pro-MMP9 *(127)*. In addition, NE exerts proinflammatory potential by enhancing neutrophil migration into inflamed tissues *(128)*. Interestingly, NE facilitates transendothelial migration of neutrophils through its ability to localize to neutrophil plasma membrane following exocytosis *(129)*.

Although initially thought to simply mediate intracellular clearance of bacteria *(130, 131)*, recent studies extend the role of NE to a modulator of inflammatory responses. Experimentally, localization of active NE to the outer surface of plasma membrane enables neutrophil transmigration in vivo *(132)*. Using *in vivo* intravital microscopy and in the presence of selective NE inhibitors, neutrophil adhesion to postcapillary venules and emigration out of vasculature were attenuated *(133)*. This role is further supported by studies using a mouse model of acute experimental arthritis that found reduced neutrophil infiltration into subsynovial tissue spaces in NE-deficient mice *(134)*, which were also found to exhibit reduced incidence of ultraviolet B (UVB) and chemically (Benzopyrene)-induced skin tumors *(87)*. It remains to be established whether reduced tumor incidence occurred as a result of impaired NE-mediated cleavage of ECM substrates, deficient MMP9 activation, or diminished cleavage of NE substrates such as ECM components or E-selectin on endothelial cells *(135)*. In addition, using an air-pouch model of inflammation, neutrophil emigration to sites of inflammation was completely attenuated in response to zymosan particles following genetic elimination of both NE and cathepsin G (a cysteine protease) *(134)*. However, neutrophil recruitment to inflamed tissues was not altered in response to lipopolysaccharide (LPS) in the absence of NE *(136)*. Notably, although neutrophils migrated normally to sites of bacterial infection in NE-deficient mice, their ability to initiate intracellular killing of gram-negative bacteria was altered *(137)*, thus indicating some degree of specificity for recruitment and response regulated by NE.

The ability of NE to differentially regulate neutrophil recruitment in "damaged" tissues, while directly significant for acute inflammatory responses, is also significant in the context of the role of neutrophils in mediating remodeling of matrix in tissues. Indeed, neutrophil-derived proteases have been identified as important regulators of insoluble elastin *(138)*, a structural component of tissues such as blood vessels, skin, and lung, in addition to hydrolysis of other ECM components, including fibronectin *(139)*, proteoglycans, and type IV collagen catabolism *(129, 134, 140)*.

Furthermore, NE can indirectly modulate structural components of tumor ECM by activating MMPs, thus facilitating tumor cell migration. Although soluble pro-MMP2 is resistant to activation by NE, pro-MMP2 becomes susceptible to activation by NE following binding to membrane-anchored MT1-MMP/MMP14 *(141)*. Furthermore, NE has also been shown to activate pro-MMP3 *(142)*, as well as pro-MMP9 *(127)*. MMP9 can in turn cleave the NE inhibitor α1-anti-trypsin, thereby indirectly enhancing NE enzymatic activity *(143)*.

Central to the proposed roles of NE in tumorigenesis is the mechanism of NE activation and identification of potential target substrates. Synthesized as an inactive zymogen, NE requires posttranslational removal of the amino-terminal dipeptide for enzymatic activation *(144)*. Through cathepsin C catalytic activity, cleavage of propeptide occurs prior or during transport of NE to neutrophil azurophil granules. Thus following neutrophil degranulation in response to various cytokines and chemoattractants, NE is secreted in its catalytically active form. One mechanism by which NE acquires resistance to the inhibitory effects of circulating proteinase inhibitors, including α1-anti-trypsin *(145)*, is by localization to neutrophil cell surface following fusion of primary granules with plasma membranes during exocytosis *(146)*. Catalytically active NE is rapidly released from azurophil granules following excessive neutrophil influx at sites of inflammation, along with terminal proteases including pro-MMP9 stored in its zymogen form in neutrophil gelatinase granules. In turn, catalytically active NE induces activation of secreted pro-MMP9 within tumor microenvironments. One way neutrophil-derived NE can significantly contribute to net tumor proteolysis is by catalyzing activation of pro-MMP9 originating from other cell populations, including infiltrated mast cells, hence resulting in amplification of MMP9-mediated effects.

2.3.4. Membrane-Anchored Serine Proteases: Emerging Roles in Cancer Development

Most serine proteases are expressed by supporting tumor stromal cells, such as mast cells and neutrophils, whereas membrane-anchored serine proteases, also known as Type II Transmembrane Serine Proteases (TTSP), appear to be largely expressed by tumor cells *(147)*. Much attention has recently been focused on membrane-anchored serine proteases, such as matriptase and hepsin, given their remarkable up-regulation in human cancers of epithelial origin, including carcinomas of skin, breast, and prostate *(148)*. Moreover, shedding of TTSP extracellular domains in seminal fluids and serum of cancer patients could represent a potential marker for cancer development and recurrence *(149)*.

Mouse models of cancer development have provided insights into which protease members of this family play significant roles in tumorigenesis processes. On the one hand, overexpression of matriptase in mouse epidermis induced spontaneous skin lesions in the absence of genetic alterations and independent of carcinogen exposure *(150)*. On the other hand, using a mouse model of non-metastasizing prostate cancer, overexpression of hepsin resulted in primary tumor progression and metastasis to liver, lung, and bone *(151)*; however, it remains to be determined how matriptase and hepsin promote neoplastic progression in these mouse models. Anchoring of TTSP to the plasma membrane directs proteolytic activities to specific compartments of tumor cell surface, including cell/ECM contacts and the invasive fronts of migrating tumor cells, thus facilitating tumor cell invasion *(151)*. Furthermore, emerging observations indicate an interesting role of matriptase and hepsin in initiating a cascade of proteolytic activations at tumor cell surface. Of note, both matriptase and hepsin have been shown to act as potent activators of receptor-bound pro-uPA *(152, 153)*, thus amplifying plasminogen conversion to plasmin by activated pro-uPA. Significantly, matriptase may promote tumor progression by directly acting as a major activator of matrix-degrading proteases, including pro-MMP3 *(154)*. In turn, activated MMP3 catalyzes activation of other MMPs, including MMP9 and MMP1 *(14)*. Extensive investigation is required to delineate upstream activators of TTSP *(148)*. Nonetheless, it is clear that tumor-associated serine proteases, notably matriptase and hepsin, contribute significantly to tumor progression via their ability to initiate a proteolytic cascade of zymogen activation, including uPA/plasmin and MMP3, collectively culminating in activation of terminal proteases, such as MMP9.

In summary, current insights regarding protease degradomics reveal that despite targeting distinct matrix substrates, serine proteases exhibit partially overlapping target substrate profiles, such as activation of pro-MMP9, whose net activity is significantly amplified during pathological processes and following simultaneous activation of various serine protease circuits originating from different cellular compartments, such as mast cells and neutrophils. Alternatively, although these protease pathways may be individually redundant, the above observations support the notion that serine proteases may profoundly influence neoplastic progression by acting collectively.

2.4. Cysteine Cathepsin Protease Regulation in Cancer Development

Cathepsins are prototypical lysosomal cysteine proteases sharing a conserved active site cleft in which amino acid residues cysteine and histidine constitute the catalytic ion pair, a distinctive characteristic of the papain-like superfamily of cysteine proteases *(155)*. Human cathepsins comprise 11 members including cathepsins B, C, H, F, K, L, O, S, V, W, and X/Z *(156)*. Although several family members have been identified as important regulators of cancer development, our laboratory has focused on cathepsin C because of the significant role it plays in regulating multiple serine proteases that together act in concert to mediate important immune-based aspects of cancer development, namely those culminating in the conversion of pro-MMP9, thus greatly amplifying MMP bioactivity.

In particular, many unique structural features contribute to the distinctive activities mediated by cathepsin C. Although most cathepsins act as endopeptidases, cathepsin C represents the only exception by acting as a dipeptidyl aminopeptidase, cleaving two-residue units from the N-terminus of a polypeptide chain *(157)*. Furthermore, localization of cathepsin C active site to the external surface of the protein confers cathepsin C with an advantage of hydrolyzing diverse groups of chymotrypsin-like proteases in their native state, regardless of size *(155)*, as opposed to other oligomeric proteases, including tryptases, where the active site is located inside of the protein *(158)*.

Cathepsin proteolytic activity is regulated at various levels, including posttranslational mechanisms. All members of the cathepsin family share in common the general mechanism of activation that is being synthesized as zymogens, therefore sharing the presence of a signal peptide and propeptide sequence removed at maturation *(159)*. To note, a residual portion of the propeptide, termed the exclusion domain, remains bound to the catalytic part of active cathepsin C *(160)* that contributes to formation and stabilization of the tetrameric structure of the mature enzyme *(161)*. *In vitro* studies exclude autocatalytic activation of pro-cathepsin C *(162)*, but whether activation of cathepsin C is facilitated by other proteases is yet to be determined.

Although initially thought to simply mediate terminal intracellular protein degradation within lysosomes, diverse biological functions of cathepsins have come to light through recent mechanistic studies, including interstitial thrombin and fibronectin metabolism, cytotoxic lymphocyte-mediated apoptotic clearance of virus-infected and tumor cells *(163)*, survival from sepsis *(164)*, and experimental arthritis *(144)*. Furthermore, using de novo carcinogenesis models in cathepsin-deficient tumor-prone mice has implicated specific roles for individual cathepsins in distinct tumorigenesis processes. A complex cascade of sequential cathepsin expression correlating with tumor development has been documented following profiling of expression and activity of cathepsins in normal, premalignant, and malignant islets of RIP1-Tag2 mice *(165, 166)*. Joyce and colleagues further assessed the unique roles of cathepsins during pancreatic islet carcinogenesis by genetic elimination of individual cathepsin genes. Although absence of cathepsin C was without consequence, tumor-associated angiogenesis was significantly reduced in the absence of cathepsin B or S, while genetic elimination of cathepsin B or L instead attenuated tumor cell proliferation and decreased tumor volume *(165)*. Similar studies by Peters and colleagues found that during mammary carcinogenesis in MMTV-PymT transgenic mice *(167)*, absence of cathepsin B emanating from macrophages significantly limited primary tumor development as well as pulmonary metastasis formation *(168)*. Interestingly, although cathepsin B was found to be a significant protumor regulator of pancreatic and mammary carcinogenesis, cathepsin B does not appear to be functionally significant during skin carcinogenesis (Junankar and Coussens, unpublished observations). Moreover, absence of cathepsin C during islet carcinogenesis is without consequence, whereas its absence during squamous carcinoma development in K14-HPV16 transgenic mice is profound (Junankar and Coussens, unpublished observations). During murine skin carcinogenesis, like in humans afflicted with loss of mutations in the *cathepsin C* gene *(169–172)*, myeloid cells fail to infiltrate damaged tissue *(144, 173, 174)*; thus, protumor programs (angiogenesis, matrix remodeling) regulated by inflammation fail to be activated.

In addition to maintaining tissue homeostasis and regulating diverse enzymatic activities, recent studies implicated cathepsin C as a mediator of inflammation. Although mice harboring homozygous deletions in the *cathepsin C* gene were resistant to experimental acute arthritis and displayed altered neutrophil recruitment in response to zymosan and immune complexes, a defective response that was rescued by administration of a neutrophil chemoattractant, cathepsin C-deficient mice exhibited normal neutrophil chemotactic responses to thioglycollate *(144)*. Likewise, on the one hand, using an air pouch model of inflammation, the number of infiltrating neutrophils was significantly attenuated in mice deficient in both NE and cathepsin G ($NE^{-/-} \times CG^{-/-}$) *(144)*, which was associated with a significant decrease in local levels of chemokines, including TNF-α and interleukin-1β (IL-1β). On the other hand, injection of IL-8, a neutrophil-specific chemokine, restored infiltration of neutrophils into air pouches of cathepsin C-deficient mice *(144)*. These results provide novel insights into the functional significance of cathepsin C-dependent proteolytic cascades in regulating local levels of chemoattractants at sites of inflammation *(144)*.

It has been recently highlighted that cathepsin C fosters tumorigenesis by acting as an important upstream regulator of multiple proteolytic events, mainly by activating several serine proteases. Using cathepsin C null homozygous mice, cathepsin C has been found to be essential for intracellular activation of neutrophil-derived proteases, including NE, cathepsin G, and proteinase 3 *(144)*. Similarly, cytotoxic T lymphocyte-derived granzymes A and B were found to be inactive and present in their proforms in the absence of cathepsin C *(174)*. Of note, while cathepsin C is necessary for activation of mouse mast cell chymase (mMCP-4) *(120)*, recent studies suggest that cathepsin C may not be essential for activation of mouse mast cell tryptases (mMCP-6) *(120, 175)*.

Indeed, *in vitro* studies support these observations, indicating that pro-β-tryptase undergoes auto-cleavage, resulting in a two amino acid residue activation dipeptide sequence that is subsequently catalyzed by cathepsin C *(176)*.

As mentioned earlier, cathepsin C initiates a cascade of proteolytic activation. Following activation by cathepsin C, mMCP-4, the major source of stored chymotrypsin-like activity in mouse peritoneum and skin *(106)*, further activates pro-MMP2 and pro-MMP9 *(107)*. In addition to mMCP-4, neutrophil-derived NE, cathepsin G, and proteinase 3 have also been found to regulate MMP2 activation *(141)*. Although these neutrophil-derived proteases are unable to process pro-MMP2 in its soluble form, recent studies indicate that pro-MMP2 undergoes conformational changes following binding to the membrane-tethered MMP14, rendering a cleavage site within pro-MMP2 prodomain region accessible for cleavage by neutrophil-derived serine proteases *(141)*. These observations raise an interesting point – that besides the notion that these proteases converge to activate common terminal proteases, redundancy should also be noted as a common theme amongst these proteolytic cascades.

In summary, using various mouse models of multistage cancer revealed interconnecting protease cascades that are initiated by a common upstream protease activator, cathepsin C, and culminating in amplification of enzymatic activity of "terminal" proteases such as MMP9 (**Fig. 2**). Significantly, these studies identified individual cathepsins that play differential roles in specific cancers emanating from multiple organ sites, and thus affirming the significance of organ and tumor type-specific regulation of protease bioactivities.

3. Intersecting Proteolytic Cascades in Cancer Development: A Target for Anti-Cancer Therapies?

Collectively, a vast body of literature indicates that proteolysis is central to neoplastic progression, in particular proteolytic enzymes emanating from host stromal cells. Given the data discussed here, it follows that proteases can potentially interact and activate other proteases in a cascade-like manner, resulting in the formation of protease circuits that may interconnect, forming the so-called *protease web (177)*. This concept may provide an alternative definition for proteolysis during tumorigenesis processes: each protease can be regarded as a "signaling molecule" that exerts its effects as part of a proteolytic pathway rather than simply functioning individually (**Fig. 1**). Nonetheless, in an attempt to understand the functional roles of proteases in cancer development, it is necessary to incorporate all the elements of the protease web, including not only proteases, but also the corresponding inhibitors, cofactors, cleaved substrates, and receptors, therefore, further viewing proteolysis as a *system*. Significantly, distinct pathological conditions may arise when perturbations occur involving key upstream proteases, e.g. cathepsin C, thus profoundly affecting a myriad of downstream effectors and ultimately altering the net proteolysis as a whole. Notably, the above-mentioned studies indicate that these linear protease circuits may converge, leading to amplification of proteolytic activity of terminal proteases like MMP9 within tissues and consequently enhancing development of pathological conditions. On the basis of these observations, and given that cathepsin C functionally contributes to neoplastic progression by acting as a key upstream proteolytic enzyme, then using a single and selective drug that targets cathepsin C proteolytic activity should impede the activation of several serine proteases, including neutrophil and mast cell-derived serine proteases, thus hold promise for effective anti-cancer therapies when compared with blocking individual serine proteases.

However, the significance of organ and tumor type-specific regulation of protease bioactivities should not be ignored when designing future anti-cancer therapies. Thus the important challenge is to characterize the "cancer degradome" at the protease and substrate levels. Which proteases are differentially active in specific tumor types? What substrates do they activate and what interconnections and networks can they potentially form? Consistent with these propositions, Joyce and colleagues investigated the effects of inhibition of cathepsin protease activity in pancreatic islet carcinogenesis using an irreversible broad-spectrum cysteine cathepsin inhibitor, JPM-OEt. Although treatment with JPM-OEt was necessary and sufficient to attenuate pancreatic tumor growth, angiogenesis, and invasiveness, however, the frequency of tumor cell apoptosis remained unaltered *(166)*. Recently, treatment with JPM-OEt in combination with the cytotoxic chemotherapeutic agent cyclophosphamide induced pancreatic tumor cell apoptosis, resulting in a more pronounced tumor regression compared with either treatment alone *(178)*. These interesting studies shed the light on the importance of using cysteine cathepsin inhibitors as effective cancer therapeutics.

Alternatively, should future studies focus on targeting multiple protease families as opposed to individual proteases? Answering this question is illustrated by recent studies documenting a more pronounced regression of pre-established tumors following simultaneous down-regulation of cathepsin B and MMP9 *(179)* as well as MMP9 and uPAR *(180, 181)* using direct intra-tumoral injections of interfering RNAs (RNAi), when compared with targeting either protease alone. Nonetheless, although these approaches are promising, future studies should take into consideration the possibility that complete loss of one enzymatic activity may be compensated for by activation of alternative pathways. For instance, following genetic elimination of cathepsin B, tumor cells have been found to induce cathepsin X expression, resulting in a partial compensation for loss of cathepsin B-mediated effects *(168)*. Taken together, the net proteolytic activity of neoplastic tissues can no longer be understood without taking into consideration the cascade of protease activities and the potential interconnections that form – that is the flow of proteolytic activities within the protease network as a whole.

4. Proteases as Therapeutic Agents

Recent studies exploited the unique localization of specific protease pathways, such as proteases that are tethered to cell surfaces, including uPA, thus directing proteolysis to discrete focal areas, versus intracellular proteolytic activities within lysosomes, such as cathepsins. These approaches have recently been used to direct activation of nontoxic "prodrugs" selectively to tumor sites *(182)*. More specifically, incorporation of a tripeptide specifier that is recognized exclusively by tumor-associated plasmin initiates release of chemotherapeutic agents, such as doxorubicin (Dox), from its prodrug form selectively in the vicinity of tumor cells, as opposed to systemic administration of Dox, which limits its activation due to the presence of physiological plasmin inhibitors, including α_2-antiplasmin and α_2-macroglobulin *(182)*. Using such novel approaches, Dox has been shown to exert its cytotoxic and/or cytostatic effects locally while restricting cardiotoxicity, a side effect that may clinically limit Dox dosage intake. It follows that similar interesting strategies have been applied to design prodrugs that are activated in tumor cells over-expressing selective protease pathways, including cathepsin B. For instance, Panchal et al. *(183)* triggered selective apoptosis of tumor cells by engineering prodrugs that are pore forming toxins, known as "prolysins." Once activated selectively by malignant tumor cells expressing high levels of membrane-associated cathepsin B, α-hemolysin is released, inducing selective permeabilization of tumor cells and ultimately, cell death.

Furthermore, recent novel advances in anti-cancer therapies took advantage of selective expression of the anti-apoptotic protein survivin in malignant ovarian cells to specifically activate expression of cytotoxic T lymphocyte-derived proapoptotic proteases, such as granzyme B *(184)*. Driven by the *survivin* promoter, active granzyme B not only reduced tumor incidence and size of xenografted human ovarian carcinoma in nude mice, but also prevented metastatic spread *(184)*. These recent studies illustrate the utility of proteases that are normally employed by immune cells to eliminate tumor cells in designing future therapeutics. Taken together, the complexity by which proteases may interact must be taken into account to delineate the physiological as well as pathological roles of proteases, as opposed to merely investigating individual proteases.

Acknowledgments

The authors acknowledge all the scientists who made contributions to the areas of research reviewed here that were not cited due to space constraints. The authors acknowledge support from the National Institutes of Health and a Department of Defense Era of Hope Scholar Award to LMC.

References

1. Bissell, M. J., Aggeler, J. (1987) Dynamic reciprocity: how do extracellular matrix and hormones direct gene expression? *Prog Clin Biol Res* 249: 251–262.

2. Bissell, M. J., Radisky, D. (2001) Putting tumours in context. *Nat Rev Cancer* 1 (1): 46–54.

3. van Kempen, L. C., de Visser, K. E., Coussens, L. M. (2006) Inflammation, proteases and cancer. *Eur J Cancer* 42 (6): 728–734.

4. Blum, G., Mullins, S. R., Keren, K., et al. (2005) Dynamic imaging of protease activity with fluorescently quenched activity-based probes. *Nat Chem Biol* 1 (4): 203–209.

5. Blum, G., von Degenfeld, G., Merchant, M. J., Blau, H. M., Bogyo, M. (2007) Noninvasive optical imaging of cysteine protease activity using fluorescently quenched activity-based probes. *Nat Chem Biol* 3 (10): 668–677.

6. Salomon, A. R., Ficarro, S. B., Brill, L. M., et al. (2003) Profiling of tyrosine phosphorylation pathways in human cells using mass spectrometry. *Proc Natl Acad Sci USA* 100 (2): 443–448.

7. Sieber, S. A., Cravatt, B. F. (2006) Analytical platforms for activity-based protein profiling-exploiting the versatility of chemistry for functional proteomics. *Chem Commun (Camb)* (22): 2311–2319.

8. Sloane, B. F., Sameni, M., Podgorski, I., Cavallo-Medved, D., Moin, K. (2006) Functional imaging of tumor proteolysis. *Annu Rev Pharmacol Toxicol* 46: 301–315.

9. Kato, D., Boatright, K. M., Berger, A. B., et al. (2005) Activity-based probes that target diverse cysteine protease families. *Nat Chem Biol* 1 (1): 33–38.

10. Lopez-Otin, C., Matrisian, L. M. (2007) Emerging roles of proteases in tumour suppression. *Nat Rev Cancer* 7 (10): 800–808.

11. Lopez-Otin, C., Overall, C. M. (2002) Protease degradomics: a new challenge for proteomics. *Nat Rev Mol Cell Biol* 3 (7): 509–519.

12. Hoffman, M. M., Monroe, D. M. (2005) Rethinking the coagulation cascade. *Curr Hematol Rep* 4 (5): 391–396.

13. Carroll, M. C. (2004) The complement system in regulation of adaptive immunity. *Nat Immunol* 5 (10): 981–986.

14. Egeblad, M., Werb, Z. (2002) New functions for the matrix metalloproteinases in cancer progression. *Nat Rev Cancer* 2: 161–174.

15. Puente, X. S., Pendás, A. M., Llano, E., Velasco, G., López-Otín, C. (1996) Molecular cloning of a novel membrane-type matrix metalloproteinase from a human breast carcinoma. *Cancer Res* 56 (5): 944–949.

16. Rhee, J. S., Coussens, L. M. (2002) RECKing MMP function: implications for cancer development. *Trends Cell Biol* 12 (5): 209–211.

17. Frank, B. T., Rossall, J. C., Caughey, G. H., Fang, K. C. (2001) Mast cell tissue inhibitor of metalloproteinase-1 is cleaved and inactivated extracellularly by alpha-chymase. *J Immunol* 166 (4): 2783–2792.

18. Coussens, L. M., Hanahan, D., Arbeit, J. M. (1996) Genetic predisposition and parameters of malignant progression in K14-HPV16 transgenic mice. *Am J Path* 149 (6): 1899–1917.

19. Coussens, L. M., Tinkle, C. L., Hanahan, D., Werb, Z. (2000) MMP-9 supplied by bone marrow-derived cells contributes to skin carcinogenesis. *Cell* 103 (3): 481–490.

20. Bergers, G., Brekken, R., McMahon, G., et al. (2000) Matrix metalloproteinase-9 triggers the angiogenic switch during carcinogenesis. *Nat Cell Biol* 2 (10): 737–744.

21. Giraudo, E., Inoue, M., Hanahan, D. (2004) An amino-bisphosphonate targets MMP-9-expressing macrophages and angiogenesis to impair cervical carcinogenesis. *J Clin Invest* 114 (5): 623–633.

22. Huang, S., Van Arsdall, M., Tedjarati, S., et al. (2002) Contributions of stromal metalloproteinase-9 to angiogenesis and growth of human ovarian carcinoma in mice. *J.Natl. Cancer Inst.* 94 (15): 1134–1142.

23. Jodele, S., Chantrain, C. F., Blavier, L., et al. (2005) The contribution of bone marrow-derived cells to the tumor vasculature in neuroblastoma is matrix metalloproteinase-9 dependent. *Cancer Res* 65 (8): 3200–3208.

24. D'Armiento, J., DiColandrea, T., Dalal, S. S., et al. (1995) Collagenase expression in transgenic mouse skin causes hyperkeratosis and acanthosis and increases susceptibility to tumorigenesis. *Mol Cell Biol* 15 (10): 5732–5739.

25. Masson, R., Lefebvre, O., Noel, A., et al. (1998) In vivo evidence that the stromelysin-3 metalloproteinase contributes in a paracrine manner to epithelial cell malignancy. *J.Cell Biol* 140 (6): 1535–1541.

26. Wilson, C. L., Heppner, K. J., Labosky, P. A., Hogan, B. L., Matrisian, L. M. (1997) Intestinal tumorigenesis is suppressed in mice lacking the metalloproteinase matrilysin. *Proc Natl Acad Sci USA* 94 (4): 1402–1407.

27. Sternlicht, M. D., Lochter, A., Sympson, C. J., et al. (1999) The stromal proteinase MMP3/stromelysin-1 promotes mammary carcinogenesis. *Cell* 98 (2): 137–146.

28. Rudolph-Owen, L. A., Cannon, P., Matrisian, L. M. (1998) Overexpression of the matrix metalloproteinase matrilysin results in premature mammary gland differentiation and male infertility. *Mol Biol Cell* 9 (2): 421–435.

29. Ha, H. Y., Moon, H. B., Nam, M. S., et al. (2001) Overexpression of membrane-type matrix metalloproteinase-1 gene induces mammary gland abnormalities and adenocarcinoma in transgenic mice. *Cancer Res* 61 (3): 984–990.

30. McCawley, L. J., Crawford, H. C., King, L. E., Jr., Mudgett, J., Matrisian, L. M. (2004) A protective role for matrix metalloproteinase-3 in squamous cell carcinoma. *Cancer Res* 64 (19): 6965–6972.

31. Balbin, M., Fueyo, A., Tester, A. M., et al. (2003) Loss of collagenase-2 confers increased skin tumor susceptibility to male mice. *Nat Genet* 35 (3): 252–257.

32. Martin, D. C., Rüther, U., Sanchez-Sweatman, O. H., Orr, F. W., Khokha, R. (1996) Inhibition of SV40 T antigen-induced hepatocellular carcinoma in TIMP-1 transgenic mice. *Oncogene* 13 (3): 569–576.

33. Buck, T. B., Yoshiji, H., Harris, S. R., Bunce, O. R., Thorgeirsson, U. P. (1999) The effects of sustained elevated levels of circulating tissue inhibitor of metalloproteinases-1 on the development of breast cancer in mice. *Ann N Y Acad Sci* 878: 732–735.

34. Rhee, J. S., Diaz, R., Korets, L., Hodgson, J. G., Coussens, L. M. (2004) TIMP-1 Alters Susceptibility to Carcinogenesis. *Cancer Res* 64 (3): 952–961.

35. Kopitz, C., Gerg, M., Bandapalli, O. R., et al. (2007) Tissue inhibitor of metalloproteinases-1 promotes liver metastasis by induction of hepatocyte growth factor signaling. *Cancer Res* 67 (18): 8615–8623.

36. Itoh, Y., Takamura, A., Ito, N., et al. (2001) Homophilic complex formation of MT1-MMP facilitates proMMP-2 activation on the cell surface and promotes tumor cell invasion. *Embo J* 20 (17): 4782–4793.

37. Wang, Z., Juttermann, R., Soloway, P. D. (2000) TIMP-2 is required for efficient activation of proMMP-2 in vivo. *J Biol Chem* 275 (34): 26411–26415.

38. Worley, J. R., Thompkins, P. B., Lee, M. H., et al. (2003) Sequence motifs of tissue inhibitor of metalloproteinases 2 (TIMP-2) determining progelatinase A (proMMP-2) binding and activation by membrane-type metalloproteinase 1 (MT1-MMP). *Biochem J* 372 (Pt 3): 799–809.

39. Bergers, G., Coussens, L. M. (2000) Extrinsic regulators of epithelial tumor progression: metalloproteinases. *Curr Opin Genet Dev* 10 (1): 120–127.

40. Chantrain, C. F., Henriet, P., Jodele, S., et al. (2006) Mechanisms of pericyte recruitment in tumour angiogenesis: a new role for metalloproteinases. *Eur J Cancer* 42 (3): 310–318.

41. Chantrain, C., Shimada, H., Jodele, S., et al. (2004) Stromal matrix metalloproteinase-9 regulates the vascular architecture in neuroblastoma by promoting pericyte recruitment. *Cancer Res* 64: 1675–1686.

42. Heissig, B., Hattori, K., Dias, S., et al. (2002) Recruitment of stem and progenitor cells from the bone marrow niche requires MMP-9 mediated release of kit-ligand. *Cell* 109 (5): 625–637.

43. Acuff, H. B., Carter, K. J., Fingleton, B., Gorden, D. L., Matrisian, L. M. (2006) Matrix metalloproteinase-9 from bone marrow-derived cells contributes to survival but not growth of tumor cells in the lung microenvironment. *Cancer Res* 66 (1): 259–266.

44. Hiratsuka, S., Nakamura, K., Iwai, S., et al. (2002) MMP9 induction by vascular endothelial growth factor receptor-1 is involved in lung-specific metastasis. *Cancer Cell* 2 (4): 289–300.

45. Giannelli, G., Falk-Marzillier, J., Schiraldi, O., Stetler-Stevenson, W. G., Quaranta, V. (1997) Induction of cell migration by matrix metalloprotease-2 cleavage of laminin-5. *Science* 277 (5323): 225–228.

46. Koshikawa, N., Giannelli, G., Cirulli, V., Miyazaki, K., Quaranta, V. (2000) Role of cell surface metalloprotease MT1-MMP in epithelial cell migration over laminin-5. *J Cell Biol* 148 (3): 615–624.

47. Lochter, A., Galosy, S., Muschler, J., Freedman, N., Werb, Z., Bissell, M. J. (1997) Matrix metalloproteinase stromelysin-1 triggers a cascade of molecular alterations that leads to stable epithelial-to-mesenchymal conversion and a premalignant phenotype in mammary epithelial cells. *J Cell Biol* 139 (7): 1861–1872.

48. Coussens, L. M., Fingleton, B., Matrisian, L. M. (2002) Matrix metalloproteinase inhibitors and cancer: trials and tribulations. *Science* 295 (5564): 2387–2392.

49. Blobel, C. P. (2005) ADAMs: key components in EGFR signalling and development. *Nat Rev Mol Cell Biol* 6 (1): 32–43.

50. Killar, L., White, J., Black, R., Peschon, J. (1999) Adamalysins. A family of metzincins including TNF-alpha converting enzyme (TACE). *Ann N Y Acad Sci* 878: 442–452.

51. White, J. M. (2003) ADAMs: modulators of cell-cell and cell-matrix interactions. *Curr Opin Cell Biol* 15 (5): 598–606.

52. Howard, L., Lu, X., Mitchell, S., Griffiths, S., Glynn, P. (1996) Molecular cloning of MADM: a catalytically active mammalian disintegrin-metalloprotease expressed in various cell types. *Biochem J* 317 (Pt 1): 45–50.

53. Howard, L., Zheng, Y., Horrocks, M., Maciewicz, R. A., Blobel, C. (2001) Catalytic activity of ADAM28. *FEBS Lett* 498 (1): 82–86.

54. Loechel, F., Gilpin, B. J., Engvall, E., Albrechtsen, R., Wewer, U. M. (1998) Human ADAM 12 (meltrin alpha) is an active metalloprotease. *J Biol Chem* 273 (27): 16993–16997.

55. Moss, M. L., Jin, S. L., Milla, M. E., et al. (1997) Cloning of a disintegrin metalloproteinase that processes precursor tumour-necrosis factor-alpha. *Nature* 385 (6618): 733–736.

56. Carl-McGrath, S., Lendeckel, U., Ebert, M., Roessner, A., Rocken, C. (2005) The disintegrin-metalloproteinases ADAM9, ADAM12, and ADAM15 are upregulated in gastric cancer. *Int J Oncol* 26 (1): 17–24.

57. Le Pabic, H., Bonnier, D., Wewer, U. M., et al. (2003) ADAM12 in human liver cancers: TGF-beta-regulated expression in stellate cells is associated with matrix remodeling. *Hepatology* 37 (5): 1056–1066.

58. Lendeckel, U., Kohl, J., Arndt, M., Carl-McGrath, S., Donat, H., Rocken, C. (2005) Increased expression of ADAM family members in human breast cancer and breast cancer cell lines. *J Cancer Res Clin Oncol* 131 (1): 41–48.

59. Peduto, L., Reuter, V. E., Shaffer, D. R., Scher, H. I., Blobel, C. P. (2005) Critical function for ADAM9 in mouse prostate cancer. *Cancer Res* 65 (20): 9312–9319.

60. Roy, R., Wewer, U. M., Zurakowski, D., Pories, S. E., Moses, M. A. (2004) ADAM 12 cleaves extracellular matrix proteins and correlates with cancer status and stage. *J Biol Chem* 279 (49): 51323–51330.

61. Wu, E., Croucher, P. I., McKie, N. (1997) Expression of members of the novel membrane linked metalloproteinase family ADAM in cells derived from a range of haematological malignancies. *Biochem Biophys Res Commun* 235 (2): 437–442.

62. Zheng, X., Jiang, F., Katakowski, M., et al. (2007) Inhibition of ADAM17 reduces hypoxia-induced brain tumor cell invasiveness. *Cancer Sci* 98 (5): 674–684.

63. Zhou, B. B., Peyton, M., He, B., et al. (2006) Targeting ADAM-mediated ligand cleavage to inhibit HER3 and EGFR pathways in non-small cell lung cancer. *Cancer Cell* 10 (1): 39–50.

64. Kenny, P. A., Bissell, M. J. (2007) Targeting TACE-dependent EGFR ligand shedding in breast cancer. *J Clin Invest* 117 (2): 337–345.

65. Evangelou, A. I., Winter, S. F., Huss, W. J., Bok, R. A., Greenberg, N. M. (2004) Steroid hormones, polypeptide growth factors, hormone refractory prostate cancer, and the neuroendocrine phenotype. *J Cell Biochem* 91 (4): 671–683.

66. Mimeault, M., Batra, S. K. (2006) Recent advances on multiple tumorigenic cascades involved in prostatic cancer progression and targeting therapies. *Carcinogenesis* 27 (1): 1–22.

67. Peduto, L., Reuter, V. E., Sehara-Fujisawa, A., Shaffer, D. R., Scher, H. I., Blobel, C. P. (2006) ADAM12 is highly expressed in

carcinoma-associated stroma and is required for mouse prostate tumor progression. *Oncogene* 25 (39): 5462–5466.

68. Kveiborg, M., Frohlich, C., Albrechtsen, R., et al. (2005) A role for ADAM12 in breast tumor progression and stromal cell apoptosis. *Cancer Res* 65 (11): 4754–4761.

69. Ringel, J., Jesnowski, R., Moniaux, N., et al. (2006) Aberrant expression of a disintegrin and metalloproteinase 17/tumor necrosis factor-alpha converting enzyme increases the malignant potential in human pancreatic ductal adenocarcinoma. *Cancer Res* 66 (18): 9045–9053.

70. Martin, J., Eynstone, L. V., Davies, M., Williams, J. D., Steadman, R. (2002) The role of ADAM 15 in glomerular mesangial cell migration. *J Biol Chem* 277 (37): 33683–33689.

71. Millichip, M. I., Dallas, D. J., Wu, E., Dale, S., McKie, N. (1998) The metallo-disintegrin ADAM10 (MADM) from bovine kidney has type IV collagenase activity in vitro. *Biochem Biophys Res Commun* 245 (2): 594–598.

72. Alfandari, D., Cousin, H., Gaultier, A., et al. (2001) Xenopus ADAM 13 is a metalloprotease required for cranial neural crest-cell migration. *Curr Biol* 11 (12): 918–930.

73. Horiuchi, K., Weskamp, G., Lum, L., et al. (2003) Potential role for ADAM15 in pathological neovascularization in mice. *Mol Cell Biol* 23 (16): 5614–5624.

74. Herren, B., Raines, E. W., Ross, R. (1997) Expression of a disintegrin-like protein in cultured human vascular cells and in vivo. *Faseb J* 11 (2): 173–180.

75. Trochon, V., Li, H., Vasse, M., et al. (1998) Endothelial metalloprotease-disintegrin protein (ADAM) is implicated in angiogenesis in vitro. *Angiogenesis* 2 (3): 277–285.

76. Yarden, Y., Sliwkowski, M. X. (2001) Untangling the ErbB signalling network. *Nat Rev Mol Cell Biol* 2 (2): 127–137.

77. Borrell-Pages, M., Rojo, F., Albanell, J., Baselga, J., Arribas, J. (2003) TACE is required for the activation of the EGFR by TGF-alpha in tumors. *Embo J* 22 (5): 1114–1124.

78. Giaccone, G., Herbst, R. S., Manegold, C., et al. (2004) Gefitinib in combination with gemcitabine and cisplatin in advanced non-small-cell lung cancer: a phase III trial--INTACT 1. *J Clin Oncol* 22 (5): 777–784.

79. Herbst, R. S., Giaccone, G., Schiller, J. H., et al. (2004) Gefitinib in combination with paclitaxel and carboplatin in advanced non-small-cell lung cancer: a phase III trial--INTACT 2. *J Clin Oncol* 22 (5): 785–794.

80. Herbst, R. S., Sandler, A. B. (2004) Overview of the current status of human epidermal growth factor receptor inhibitors in lung cancer. *Clin Lung Cancer* 6 Suppl 1: S7–S19.

81. Huovila, A. P., Turner, A. J., Pelto-Huikko, M., Karkkainen, I., Ortiz, R. M. (2005) Shedding light on ADAM metalloproteinases. *Trends Biochem Sci* 30 (7): 413–422.

82. Mohammed, F. F., Smookler, D. S., Taylor, S. E., et al. (2004) Abnormal TNF activity in Timp3-/- mice leads to chronic hepatic inflammation and failure of liver regeneration. *Nat Genet* 36 (9): 969–977.

83. Toth, M., Sohail, A., Mobashery, S., Fridman, R. (2006) MT1-MMP shedding involves an ADAM and is independent of its localization in lipid rafts. *Biochem Biophys Res Commun* 350 (2): 377–384.

84. Osenkowski, P., Toth, M., Fridman, R. (2004) Processing, shedding, and endocytosis of membrane type 1-matrix metalloproteinase (MT1-MMP). *J Cell Physiol* 200 (1): 2–10.

85. Bugge, T. H., Lund, L. R., Kombrinck, K. K., et al. (1998) Reduced metastasis of Polyoma virus middle T antigen-induced mammary cancer in plasminogen-deficient mice. *Oncogene* 16 (24): 3097–3104.

86. Coussens, L. M., Raymond, W. W., Bergers, G., et al. (1999) Inflammatory mast cells upregulate angiogenesis during squamous epithelial carcinogenesis. *Genes Dev* 13 (11): 1382–1397.

87. Starcher, B., O'Neal, P., Granstein, R. D., Beissert, S. (1996) Inhibition of neutrophil elastase suppresses the development of skin tumors in hairless mice. *J Invest Dermatol* 107 (2): 159–163.

88. Netzel-Arnett, S., Hooper, J. D., Szabo, R., et al. (2003) Membrane anchored serine proteases: a rapidly expanding group of cell surface proteolytic enzymes with potential roles in cancer. *Cancer Metastasis Rev* 22 (2–3): 237–258.

89. Caughey, G. H. (2002) New developments in the genetics and activation of mast cell proteases. *Mol Immunol* 38 (16–18): 1353–1357.

90. Reiling, K. K., Krucinski, J., Miercke, L. J., Raymond, W. W., Caughey, G. H., Stroud, R. M. (2003) Structure of human pro-chymase: a model for the activating transition of granule-associated proteases. *Biochemistry* 42 (9): 2616–2624.

91. Ramos-DeSimone, N., Hahn-Dantona, E., Sipley, J., Nagase, H., French, D. L., Quigley, J. P. (1999) Activation of matrix metalloproteinase-9 (MMP-9) via a converging plasmin/stromelysin-1 cascade enhances tumor cell invasion. *J Biol Chem* 274 (19): 13066–13076.

92. Pollanen, J., Hedman, K., Nielsen, L. S., Dano, K., Vaheri, A. (1988) Ultrastructural localization of plasma membrane-associated urokinase-type plasminogen activator at focal contacts. *J Cell Biol* 106 (1): 87–95.

93. Ellis, V., Behrendt, N., Dano, K. (1991) Plasminogen activation by receptor-bound urokinase. A kinetic study with both cell-associated and isolated receptor. *J Biol Chem* 266 (19): 12752–12758.

94. Wun, T. C., Reich, E. (1987) An inhibitor of plasminogen activation from human placenta. Purification and characterization. *J Biol Chem* 262 (8): 3646–3653.

95. Conese, M., Blasi, F. (1995) Urokinase/ urokinase receptor system: internalization/ degradation of urokinase-serpin complexes: mechanism and regulation. *Biol Chem Hoppe Seyler* 376 (3): 143–155.

96. Ploplis, V., Tipton, H., Menchen, H., Castellino, F. (2007) A urokinase-type plasminogen activator deficiency diminishes the frequency of intestinal adenomas in Apc(Min/+) mice. *J Pathol* 213 (3): 266–274.

97. Ploplis, V. A. (2001) Gene targeting in hemostasis. plasminogen. *Front Biosci* 6: D555–D569.

98. Hahn-Dantona, E., Ramos-DeSimone, N., Sipley, J., Nagase, H., French, D. L., Quigley, J. P. (1999) Activation of proMMP-9 by a plasmin/MMP-3 cascade in a tumor cell model. Regulation by tissue inhibitors of metalloproteinases. *Ann N Y Acad Sci* 878: 372–387.

99. He, C. S., Wilhelm, S. M., Pentland, A. P., et al. (1989) Tissue cooperation in a proteolytic cascade activating human interstitial collagenase. *Proc Natl Acad Sci USA* 86 (8): 2632–2636.

100. Monea, S., Lehti, K., Keski-Oja, J., Mignatti, P. (2002) Plasmin activates pro-matrix metalloproteinase-2 with a membrane-type 1 matrix metalloproteinase-dependent mechanism. *J Cell Physiol* 192 (2): 160–170.

101. Blank, U., Rivera, J. (2004) The ins and outs of IgE-dependent mast-cell exocytosis. *Trends Immunol* 25 (5): 266–273.

102. el-Lati, S. G., Dahinden, C. A., Church, M. K. (1994) Complement peptides C3a- and C5a-induced mediator release from dissociated human skin mast cells. *J Invest Dermatol* 102 (5): 803–806.

103. Karimi, K., Redegeld, F. A., Blom, R., Nijkamp, F. P. (2000) Stem cell factor and interleukin-4 increase responsiveness of mast cells to substance P. *Exp Hematol* 28 (6): 626–634.

104. Hogaboam, C., Kunkel, S. L., Strieter, R. M., et al. (1998) Novel role of transmembrane SCF for mast cell activation and eotaxin production in mast cell-fibroblast interactions. *J Immunol* 160 (12): 6166–6171.

105. Kulka, M., Alexopoulou, L., Flavell, R. A., Metcalfe, D. D. (2004) Activation of mast cells by double-stranded RNA: evidence for activation through Toll-like receptor 3. *J Allergy Clin Immunol* 114 (1): 174–182.

106. Tchougounova, E., Pejler, G., Abrink, M. (2003) The chymase, mouse mast cell protease 4, constitutes the major chymotrypsin-like activity in peritoneum and ear tissue. A role for mouse mast cell protease 4 in thrombin regulation and fibronectin turnover. *J Exp Med* 198 (3): 423–431.

107. Tchougounova, E., Lundequist, A., Fajardo, I., Winberg, J. O., Abrink, M., Pejler, G. (2005) A key role for mast cell chymase in the activation of pro-matrix metalloprotease-9 and pro-matrix metalloprotease-2. *J Biol Chem* 280 (10): 9291–9296.

108. Ghildyal, N., Friend, D. S., Stevens, R. L., et al. (1996) Fate of two mast cell tryptases in V3 mastocytosis and normal BALB/c mice undergoing passive systemic anaphylaxis: prolonged retention of exocytosed mMCP-6 in connective tissues, and rapid accumulation of enzymatically active mMCP-7 in the blood. *J Exp Med* 184 (3): 1061–1073.

109. Wong, G. W., Yasuda, S., Morokawa, N., Li, L., Stevens, R. L. (2004) Mouse chromosome 17A3.3 contains 13 genes that encode functional tryptic-like serine proteases with distinct tissue and cell expression patterns. *J Biol Chem* 279 (4): 2438–2452.

110. Caughey, G. H. (2007) Mast cell tryptases and chymases in inflammation and host defense. *Immunol Rev* 217: 141–154.

111. Schechter, I., Berger, A. (1967) On the size of the active site in proteases. I. Papain. *Biochem Biophys Res Commun* 27 (2): 157–162.

112. Schechter, N. M., Choi, J. K., Slavin, D. A., et al. (1986) Identification of a chymotrypsin-like proteinase in human mast cells. *J Immunol* 137 (3): 962–970.

113. He, S., Walls, A. F. (1998) Human mast cell chymase induces the accumulation of neutrophils, eosinophils and other inflammatory cells in vivo. *Br J Pharmacol* 125 (7): 1491–1500.

114. Gruber, B. L., Marchese, M. J., Suzuki, K., et al. (1989) Synovial procollagenase activation by human mast cell tryptase dependence upon matrix metalloproteinase 3 activation. *J Clin Invest* 84 (5): 1657–1662.

115. Hartmann, T., Ruoss, S. J., Raymond, W. W., Seuwen, K., Caughey, G. H. (1992) Human tryptase as a potent, cell-specific mitogen: role of signaling pathways in synergistic responses. *Am J Physiol* 262 (5 Pt 1): L528–L534.

116. Ruoss, S. J., Hartmann, T., Caughey, G. H. (1991) Mast cell tryptase is a mitogen for cultured fibroblasts. *J Clin Invest* 88 (2): 493–499.

117. Cairns, J. A., Walls, A. F. (1996) Mast cell tryptase is a mitogen for epithelial cells. Stimulation of IL-8 production and intercellular adhesion molecule-1 expression. *J Immunol* 156 (1): 275–283.

118. Huang, C., Friend, D. S., Qiu, W. T., et al. (1998) Induction of a selective and persistent extravasation of neutrophils into the peritoneal cavity by tryptase mouse mast cell protease 6. *J Immunol* 160 (4): 1910–1919.

119. Compton, S. J., Cairns, J. A., Holgate, S. T., Walls, A. F. (1999) Interaction of human mast cell tryptase with endothelial cells to stimulate inflammatory cell recruitment. *Int Arch Allergy Immunol* 118 (2–4): 204–205.

120. Wolters, P. J., Pham, C. T., Muilenburg, D. J., Ley, T. J., Caughey, G. H. (2001) Dipeptidyl peptidase I is essential for activation of mast cell chymases, but not tryptases, in mice. *J Biol Chem* 276 (21): 18551–18566.

121. Fouret, P., du Bois, R. M., Bernaudin, J. F., Takahashi, H., Ferrans, V. J., Crystal, R. G. (1989) Expression of the neutrophil elastase gene during human bone marrow cell differentiation. *J Exp Med* 169 (3): 833–845.

122. Zimmer, M., Medcalf, R. L., Fink, T. M., Mattmann, C., Lichter, P., Jenne, D. E. (1992) Three human elastase-like genes coordinately expressed in the myelomonocyte lineage are organized as a single genetic locus on 19pter. *Proc Natl Acad Sci USA* 89 (17): 8215–8219.

123. Chua, F., Laurent, G. J. (2006) Neutrophil elastase: mediator of extracellular matrix destruction and accumulation. *Proc Am Thorac Soc* 3 (5): 424–427.

124. Akizuki, M., Fukutomi, T., Takasugi, M., et al. (2007) Prognostic significance of immunoreactive neutrophil elastase in human breast cancer: long-term follow-up results in 313 patients. *Neoplasia* 9 (3): 260–264.

125. Yamashita, J., Tashiro, K., Yoneda, S., Kawahara, K., Shirakusa, T. (1996) Local increase in polymorphonuclear leukocyte elastase is associated with tumor invasiveness in non-small cell lung cancer. *Chest* 109 (5): 1328–1334.

126. Lane, A., Ley, T. J. (2003) Neutrophil elastase cleaves PML-RARalpha and is important for the development of acute promyelocytic leukemia in mice. *Cell* 1115: 305–318.

127. Ferry, G., Lonchampt, M., Pennel, L., de Nanteuil, G., Canet, E., Tucker, G. C. (1997) Activation of MMP-9 by neutrophil elastase in an in vivo model of acute lung injury. *FEBS Lett* 402 (2–3): 111–115.

128. Nakamura, H., Yoshimura, K., McElvaney, N. G., Crystal, R. G. (1992) Neutrophil elastase in respiratory epithelial lining fluid of individuals with cystic fibrosis induces interleukin-8 gene expression in a human bronchial epithelial cell line. *J Clin Invest* 89 (5): 1478–1484.

129. Cepinskas, G., Sandig, M., Kvietys, P. R. (1999) PAF-induced elastase-dependent neutrophil transendothelial migration is associated with the mobilization of elastase to the neutrophil surface and localization to the migrating front. *J Cell Sci* 112 (Pt 12): 1937–1945.

130. Belaaouaj, A., McCarthy, R., Baumann, M., et al. (1998) Mice lacking neutrophil elastase reveal impaired host defense against gram negative bacterial sepsis. *Nat Med* 4 (5): 615–618.

131. Brinkmann, V., Reichard, U., Goosmann, C., et al. (2004) Neutrophil extracellular traps kill bacteria. *Science* 303 (5663): 1532–1535.

132. Young, R. E., Voisin, M. B., Wang, S., Dangerfield, J., Nourshargh, S. (2007) Role of neutrophil elastase in LTB(4)-induced neutrophil transmigration in vivo assessed with a specific inhibitor and neutrophil elastase deficient mice. *Br J Pharmacol* 151 (5): 628–637.

133. Woodman, R. C., Reinhardt, P. H., Kanwar, S., Johnston, F. L., Kubes, P. (1993) Effects of human neutrophil elastase (HNE) on neutrophil function in vitro and in inflamed microvessels. *Blood* 82 (7): 2188–2195.

134. Sato, T., Takahashi, S., Mizumoto, T., et al. (2006) Neutrophil elastase and cancer. *Surg Oncol* 15 (4): 217–222.

135. Nozawa, F., Hirota, M., Okabe, A., et al. (2000) Elastase activity enhances the adhesion of neutrophil and cancer cells to vascular endothelial cells. *J Surg Res* 94 (2): 153–158.

136. Hirche, T. O., Atkinson, J. J., Bahr, S., Belaaouaj, A. (2004) Deficiency in neutrophil elastase does not impair neutrophil recruitment to inflamed sites. *Am J Respir Cell Mol Biol* 30 (4): 576–584.

137. Young, R. E., Thompson, R. D., Larbi, K. Y., et al. (2004) Neutrophil elastase (NE)-deficient mice demonstrate a nonredundant role

for NE in neutrophil migration, generation of proinflammatory mediators, and phagocytosis in response to zymosan particles in vivo. *J Immunol* 172 (7): 4493–4502.

138. Baugh, R. J., Travis, J. (1976) Human leukocyte granule elastase: rapid isolation and characterization. *Biochemistry* 15 (4): 836–841.

139. McDonald, J. A., Kelley, D. G. (1980) Degradation of fibronectin by human leukocyte elastase. Release of biologically active fragments. *J Biol Chem* 255 (18): 8848–8858.

140. Mainardi, C. L., Dixit, S. N., Kang, A. H. (1980) Degradation of type IV (basement membrane) collagen by a proteinase isolated from human polymorphonuclear leukocyte granules. *J Biol Chem* 255 (11): 5435–5441.

141. Shamamian, P., Schwartz, J. D., Pocock, B. J., et al. (2001) Activation of progelatinase A (MMP-2) by neutrophil elastase, cathepsin G, and proteinase-3: a role for inflammatory cells in tumor invasion and angiogenesis. *J Cell Physiol* 189 (2): 197–206.

142. Okada, Y., Nakanishi, I. (1989) Activation of matrix metalloproteinase 3 (stromelysin) and matrix metalloproteinase 2 ('gelatinase') by human neutrophil elastase and cathepsin G. *FEBS Lett.* 249 (2): 353–356.

143. Liu, Z., Zhou, X., Shapiro, S. D., et al. (2000) The serpin alpha1-proteinase inhibitor is a critical substrate for gelatinase B/MMP-9 in vivo. *Cell* 102 (5): 647–655.

144. Adkison, A. M., Raptis, S. Z., Kelley, D. G., Pham, C. T. (2002) Dipeptidyl peptidase I activates neutrophil-derived serine proteases and regulates the development of acute experimental arthritis. *J Clin Invest* 109 (3): 363–371.

145. Owen, C. A., Campbell, M. A., Sannes, P. L., Boukedes, S. S., Campbell, E. J. (1995) Cell surface-bound elastase and cathepsin G on human neutrophils: a novel, non-oxidative mechanism by which neutrophils focus and preserve catalytic activity of serine proteinases. *J Cell Biol* 131 (3): 775–789.

146. Lee, W. L., Downey, G. P. (2001) Leukocyte elastase: physiological functions and role in acute lung injury. *Am J Respir Crit Care Med* 164 (5): 896–904.

147. Matrisian, L. M. (1999) Cancer biology: extracellular proteinases in malignancy. *Curr Biol* 9 (20): R776–R778.

148. List, K., Bugge, T. H., Szabo, R. (2006) Matriptase: potent proteolysis on the cell surface. *Mol Med* 12 (1–3): 1–7.

149. Mok, S. C., Chao, J., Skates, S., et al. (2001) Prostasin, a potential serum marker for ovarian cancer: identification through microarray technology. *J Natl Cancer Inst* 93 (19): 1458–1464.

150. List, K., Szabo, R., Molinolo, A., et al. (2005) Deregulated matriptase causes ras-independent multistage carcinogenesis and promotes ras-mediated malignant transformation. *Genes Dev* 19 (16): 1934–1950.

151. Klezovitch, O., Chevillet, J., Mirosevich, J., Roberts, R. L., Matusik, R. J., Vasioukhin, V. (2004) Hepsin promotes prostate cancer progression and metastasis. *Cancer Cell* 6 (2): 185–195.

152. Moran, P., Li, W., Fan, B., Vij, R., Eigenbrot, C., Kirchhofer, D. (2006) Pro-urokinase-type plasminogen activator is a substrate for hepsin. *J Biol Chem* 281 (41): 30439–30446.

153. Suzuki, M., Kobayashi, H., Kanayama, N., et al. (2004) Inhibition of tumor invasion by genomic down-regulation of matriptase through suppression of activation of receptor-bound pro-urokinase. *J Biol Chem* 279 (15): 14899–14908.

154. Jin, X., Yagi, M., Akiyama, N., et al. (2006) Matriptase activates stromelysin (MMP-3) and promotes tumor growth and angiogenesis. *Cancer Sci* 97 (12): 1327–1334.

155. Turk, D., Janjic, V., Stern, I., et al. (2001) Structure of human dipeptidyl peptidase I (cathepsin C): exclusion domain added to an endopeptidase framework creates the machine for activation of granular serine proteases. *Embo J* 20 (23): 6570–6582.

156. Turk, V., Turk, B., Guncar, G., Turk, D., Kos, J. (2002) Lysosomal cathepsins: structure, role in antigen processing and presentation, and cancer. *Adv Enzyme Regul* 42: 285–303.

157. McGuire, M. J., Lipsky, P. E., Thiele, D. L. (1992) Purification and characterization of dipeptidyl peptidase I from human spleen. *Arch Biochem Biophys* 295 (2): 280–288.

158. Hallgren, J., Pejler, G. (2006) Biology of mast cell tryptase. An inflammatory mediator. *Febs J* 273 (9): 1871–1895.

159. Vasiljeva, O., Reinheckel, T., Peters, C., Turk, D., Turk, V., Turk, B. (2007) Emerging roles of cysteine cathepsins in disease and their potential as drug targets. *Curr Pharm Des* 13 (4): 387–403.

160. Dolenc, I., Turk, B., Pungercic, G., Ritonja, A., Turk, V. (1995) Oligomeric structure and substrate induced inhibition of human cathepsin C. *J Biol Chem* 270 (37): 21626–21631.

161. Cigic, B., Dahl, S. W., Pain, R. H. (2000) The residual pro-part of cathepsin C fulfills

the criteria required for an intramolecular chaperone in folding and stabilizing the human proenzyme. *Biochemistry* 39 (40): 12382–12390.

162. Dahl, S. W., Halkier, T., Lauritzen, C., et al. (2001) Human recombinant pro-dipeptidyl peptidase I (cathepsin C) can be activated by cathepsins L and S but not by autocatalytic processing. *Biochemistry* 40 (6): 1671–1678.

163. Shresta, S., Pham, C. T., Thomas, D. A., Graubert, T. A., Ley, T. J. (1998) How do cytotoxic lymphocytes kill their targets? *Curr Opin Immunol* 10 (5): 581–587.

164. Mallen-St Clair, J., Pham, C. T., Villalta, S. A., Caughey, G. H., Wolters, P. J. (2004) Mast cell dipeptidyl peptidase I mediates survival from sepsis. *J Clin Invest* 113 (4): 628–634.

165. Gocheva, V., Zeng, W., Ke, D., et al. (2006) Distinct roles for cysteine cathepsin genes in multistage tumorigenesis. *Genes Dev* 20 (5): 543–556.

166. Joyce, J. A., Baruch, A., Chehade, K., et al. (2004) Cathepsin cysteine proteases are effectors of invasive growth and angiogenesis during multistage tumorigenesis. *Cancer Cell* 5 (5): 443–453.

167. Guy, C. T., Cardiff, R. D., Muller, W. J. (1992) Induction of mammary tumors by expression of polyomavirus middle T oncogene: a transgenic mouse model for metastatic disease. *Mol Cell Biol* 12 (3): 954–961.

168. Vasiljeva, O., Papazoglou, A., Kruger, A., et al. (2006) Tumor cell-derived and macrophage-derived cathepsin B promotes progression and lung metastasis of mammary cancer. *Cancer Res* 66 (10): 5242–5250.

169. de Haar, S. F., Jansen, D. C., Schoenmaker, T., De Vree, H., Everts, V., Beertsen, W. (2004) Loss-of-function mutations in cathepsin C in two families with Papillon-Lefevre syndrome are associated with deficiency of serine proteinases in PMNs. *Hum Mutat* 23 (5): 524.

170. Frezzini, C., Leao, J. C., Porter, S. (2004) Cathepsin C involvement in the aetiology of Papillon-Lefevre syndrome. *Int J Paediatr Dent* 14 (6): 466–467.

171. Hewitt, C., McCormick, D., Linden, G., et al. (2004) The role of cathepsin C in Papillon-Lefevre syndrome, prepubertal periodontitis, and aggressive periodontitis. *Hum Mutat* 23 (3): 222–228.

172. Noack, B., Gorgens, H., Hoffmann, T., et al. (2004) Novel mutations in the cathepsin C gene in patients with pre-pubertal aggressive periodontitis and Papillon-Lefevre syndrome. *J Dent Res* 83 (5): 368–370.

173. Pham, C. T., Ivanovich, J. L., Raptis, S. Z., Zehnbauer, B., Ley, T. J. (2004) Papillon-Lefevre syndrome: correlating the molecular, cellular, and clinical consequences of cathepsin C/dipeptidyl peptidase I deficiency in humans. *J Immunol* 173 (12): 7277–7281.

174. Pham, C. T., Ley, T. J. (1999) Dipeptidyl peptidase I is required for the processing and activation of granzymes A and B in vivo. *Proc Natl Acad Sci USA* 96 (15): 8627–8632.

175. Sheth, P. D., Pedersen, J., Walls, A. F., McEuen, A. R. (2003) Inhibition of dipeptidyl peptidase I in the human mast cell line HMC-1: blocked activation of tryptase, but not of the predominant chymotryptic activity. *Biochem Pharmacol* 66 (11): 2251–2262.

176. Sakai, K., Ren, S., Schwartz, L. B. (1996) A novel heparin-dependent processing pathway for human tryptase. Autocatalysis followed by activation with dipeptidyl peptidase I. *J Clin Invest* 97 (4): 988–995.

177. Overall, C. M., Kleifeld, O. (2006) Tumour microenvironment - opinion: validating matrix metalloproteinases as drug targets and anti-targets for cancer therapy. *Nat Rev Cancer* 6 (3): 227–239.

178. Bell-McGuinn, K. M., Garfall, A. L., Bogyo, M., Hanahan, D., Joyce, J. A. (2007) Inhibition of cysteine cathepsin protease activity enhances chemotherapy regimens by decreasing tumor growth and invasiveness in a mouse model of multistage cancer. *Cancer Res* 67 (15): 7378–7385.

179. Lakka, S. S., Gondi, C. S., Yanamandra, N., et al. (2004) Inhibition of cathepsin B and MMP-9 gene expression in glioblastoma cell line via RNA interference reduces tumor cell invasion, tumor growth and angiogenesis. *Oncogene* 23 (27): 4681–4689.

180. Lakka, S. S., Gondi, C. S., Dinh, D. H., et al. (2005) Specific interference of urokinase-type plasminogen activator receptor and matrix metalloproteinase-9 gene expression induced by double-stranded RNA results in decreased invasion, tumor growth, and angiogenesis in gliomas. *J Biol Chem* 280 (23): 21882–21892.

181. Lakka, S. S., Gondi, C. S., Yanamandra, N., et al. (2003) Synergistic down-regulation of urokinase plasminogen activator receptor and matrix metalloproteinase-9 in SNB19 glioblastoma cells efficiently inhibits glioma cell invasion, angiogenesis, and tumor growth. *Cancer Res* 63 (10): 2454–2461.

182. Devy, L., de Groot, F. M., Blacher, S., et al. (2004) Plasmin-activated doxorubicin prodrugs containing a spacer reduce tumor growth and angiogenesis without systemic toxicity. *Faseb J* 18 (3): 565–567.

183. Panchal, R. G., Cusack, E., Cheley, S., Bayley, H. (1996) Tumor protease-activated, pore-forming toxins from a combinatorial library. *Nat Biotechnol* 14 (7): 852–856.

184. Caldas, H., Jaynes, F. O., Boyer, M. W., Hammond, S., Altura, R. A. (2006) Survivin and Granzyme B-induced apoptosis, a novel anticancer therapy. *Mol Cancer Ther* 5 (3): 693–703.

185. Jodele, S., Chantrain C. F., Blavier, L., Lutzko, C., Crooks, G. M., Shimada, H., Coussens, L. M., Declerck, Y. A. (2005) The contribution of bone marrow-derived cells to the tumor vasculature in neuroblastoma is matrix metalloproteinase-9 dependent. Cancer Res 65 (8): 3200–3208

Chapter 2

Proteolytic Systems: Constructing Degradomes

Gonzalo R. Ordóñez, Xose S. Puente,
Víctor Quesada, and Carlos López-Otín

Summary

Proteolytic enzymes play an essential role in many biological and pathological processes. Taking advantage of the recent availability of several mammalian genome sequences and by using a set of computational approaches, we have annotated and compared the degradome or complete repertoire of proteases of different mammalian species including human, mouse, rat, and chimpanzee. These studies have allowed us to expand our knowledge about the complexity, evolution, and diversity of proteolytic systems, which represent about 2% of the studied genomes. In this chapter, we review the genomic and computational methodologies used in this degradomic analysis and summarize the main findings derived from comparison of mammalian degradomes.

Key words: Protease, Degradome, Genome, Evolution, Disease, Bioinformatics.

1. Protease Genomics and the Degradome

Proteases are enzymes with the common ability to hydrolyze peptide bonds through nucleophilic attack on the carbonyl group by a protein residue or a polarized water molecule. According to the hydrolysis mechanism and the involved group, proteases can be classified in six different groups. In serine, cysteine, and threonine proteases, the catalytic nucleophile is a hydroxyl (serine and threonine) or sulfhydryl (cysteine) group of the corresponding side chain. In aspartic, glutamic, and metalloproteases, an activated water molecule acts as a nucleophile to attack the peptide bond of the substrate (1–4).

Proteases were initially described in nonspecific reactions of protein catabolism such as degradation of dietary proteins or tissue destruction. However, this concept has been widely revised

Toni M. Antalis and Thomas H. Bugge (eds.), *Methods in Molecular Biology, Proteases and Cancer, Vol. 539*
© Humana Press, a part of Springer Science + Business Media, LLC 2009
DOI: 10.1007/978-1-60327-003-8_2

and it is now well established that proteases take part in highly selective and limited cleavage of specific substrates. Hence, proteolytic enzymes regulate many important cellular functions and have fundamental roles in biological processes, such as cell cycle progression, cell proliferation, differentiation and migration, embryonic development, tissue remodeling, angiogenesis, apoptosis, autophagy, senescence, fertilization, blood coagulation, immunity, wound healing, or hemostasis (5, 6).

Due to the biological relevance of proteases, deficiencies or changes in their spatial and temporal regulation are the cause for severe human diseases such as cancer, arthritis, and neurodegenerative and cardiovascular disorders (7–9). Moreover, at least 74 human hereditary diseases are caused by mutations in protease-coding genes owing to protease loss or gain of function. Some examples are hemophilia (10), Alzheimer disease (11), hereditary pancreatitis (12), and progeroid syndromes (13) caused by mutations in coagulation factors, presenilins, cationic trypsinogen, or FACE-1 metalloproteinase, respectively.

The growing importance of proteases in human biology and pathology together with their diversity and increasing relevance as therapeutic targets has made necessary the use of novel concepts for the global study of proteolysis (14). Thus, the term degradome was defined as the complete set of proteases that are expressed at a specific moment or circumstance by a cell, tissue, or organism. Likewise, the degradome of a protease should be the repertoire of substrates targeted by that protease. Finally, degradomics comprises all genomic and proteomic approaches for the identification and characterization of proteases that are present in an organism, including the substrates that are targeted by these proteases and their endogenous inhibitors (14).

The recent availability of the complete genome sequences of a number of organisms, including several mammalian species (15–19), has raised the possibility of studying the degradome in depth, by facilitating the definition of the complete repertoire of protease coding genes in a genome as well as the identification and analysis of novel proteases. Furthermore, the study of specific protease gene families has proven very useful to facilitate the analysis of the large amount of information included in a genome and to understand the mechanisms implicated in the evolution of mammalian organisms.

2. Tools for Degradome Analysis

Genome annotation can be defined as the part of the genome analysis that identifies elements on the genome and relates them to biological information (20). This task requires a certain level of automation because of the size of a genome and the large number of elements that can be found in it *(21, 22)*. This automation is especially important to run routine tasks and to organize the results. However, although automated annotation is clearly desirable, the lack of significant experimental information in many sequenced genomes as that available in human or other model organisms implies that manual annotation remains the gold standard in eukaryotic genomes analysis. Thus, coding sequence, distribution of exons and introns, and other gene features must be analyzed individually for every gene by expert scientists in order to assure its quality and reliability. Currently, even the best computational methods for gene prediction are far from properly predict the complete set of genes in a genome *(23)*. Due to this limitation, this chapter will be mainly focused on the manual annotation of genomes (and more specifically degradomes) using comparison methodologies.

Prokaryotic genes are much easier to annotate than eukaryotic ones because they can be defined as the longest open reading frame (ORF) for a given region of DNA. Translation of DNA genomic sequence in all six reading frames is a relatively simple task *(24)*. Moreover, GC content and presence of a ribosome-binding site or a promoter are additional evidences for a coding gene. Nevertheless, multicellular eukaryotes, especially mammals, have huge genomic sequences with large intergenic regions, and a complex gene organization as genes are discontinuous with coding exons separated by noncoding introns. Moreover, the transcribed mRNA also contains untranslated regions upstream from the first exon (5′-UTR) and downstream from the last exon (3′-UTR). Furthermore, for many genes there are multiple expressed forms due to alternative splicing processes. Thus, gene identification turns into a major problem. Gene prediction approaches should identify all exons, even those in 5′-UTR and 3′-UTR including introns and alternative splicing forms. However, for practical purposes, it is sufficient to identify and properly annotate translated exons to deduce the sequence of the corresponding protein.

Only a small fraction of all known proteins have been functionally annotated. Therefore, the use of bioinformatics approaches to sequence analysis for the annotation of novel genes or genomes and the prediction of protein function and structure is critical. This identification and annotation of coding sequences in the DNA is still one of the most imperative problems in genome analysis. Several prediction methodologies have been developed for this task. Some of them perform gene prediction ab initio, in other words, they only use statistical parameters of the DNA sequence for gene identification *(25–27)*. Conversely, there are homology-based methods which identify novel sequences in genomes or databases by comparison algorithms with previously known or predicted sequences. Before discussing these methods and algorithms, it is necessary to define several important concepts that will be used in next pages.

We must first emphasize that the main goal of DNA and protein sequence analysis and comparison is to identify *homologous sequences* and once identified, try to predict common activities, biological functions, or evolutionary relationships. To address this objective, it is important to distinguish different types of homologous relationships. The two categories are *orthologs*, defined as homologous genes that have originated as a result of a speciation event, derived from a single ancestral gene in the last common ancestor of two given species, and *paralogs*, which are homologous genes that have originated as a result of a duplication event within the same genome *(28)*. Orthologous genes typically retain the same ancestral function, while parologous genes tend to evolve to new functions. When paralogs arise owing to gene duplication events, the pressure of purifying selection decreases for one or both paralogs *(29, 30)*. This fact allows the evolution of new functions. Therefore, identification of orthologous gene sets in complete genomes can be a difficult task because of the occurrence of multiple duplications, speciation, or lineage-specific gene loss or expansion events.

When a wide-genome search is performed using a comparative method, amino acid sequence comparisons have multiple advantages over nucleotide sequence comparisons, which highly improve sensitivity. There are 20 amino acids but only four bases, so an amino acid match results in more information than a nucleotide match. Furthermore, because of genetic code degeneration, almost one-third of nucleotides in coding sequences represent noise. In addition, nucleotide databases are much larger and less informative than protein databases due to the great amounts of noncoding sequences that are deposited in databases by genome sequencing centers. Finally, protein comparison methods take advantage of the fact that, under the principle of parsimony, related proteins are likely to have similar residues at any position that is not perfectly conserved. In practice, this additional feature means that weak but significant similarities will be better detected by protein comparison than by nucleotide comparison methods. Because of these advantages, identification of coding sequences is usually performed using protein sequences, although the final objective is to obtain a DNA alignment. Nucleotide sequence comparisons are required only when noncoding regions (5′-UTRs, 3′-UTRs, introns, or promoters) are analyzed.

Each of the 20 amino acids has unique properties. Accordingly, its substitution by another amino acid will affect the protein's structure and function. In general, the more similar the properties of two different residues, the less will be the effect. Hence, this kind of conservative substitutions should be penalized less than a major change in sequence comparisons. To measure the likelihood of a substitution between residues i and j, the following formula can be used: $S_{ij} = k \ln(q_{ij}/p_i p_j)$, where k is a coefficient, q_{ij} is the frequency of the substitution and p_i and p_j are the frequencies of the respective residues calculated from alignments of homologous protein families. With these likelihoods (scores) for every pair of amino acid substitution, a *substitution score matrix* (SSM) can be built *(31)*. These SSMs play an essential role in computational methods used to detect similarities among sequences. The most widely used groups of matrices are the BLOSUM series (BLOcks SUbstitution Matrix) *(32)*. For example, BLOSUM62 matrix substitution scores were derived from aligning sequences with less than 62% identity, and it is used as the default in most database searches.

Another issue that can hinder gene prediction is the presence of sequencing errors in the analyzed sequence. Error correction techniques should be used with caution because eukaryotic genomes contain numerous *pseudogenes*, and frameshift correction might annotate pseudogenes as functional genes. A pseudogene is a nonfunctional copy of a gene. There are two main types of pseudogenes. A *conventional pseudogene* is a gene that has been inactivated by mutation. Once a pseudogene has become nonfunctional it will degenerate through accumulation of more mutations and eventually will no longer be recognizable as a gene. A *processed pseudogene* is derived from the mRNA copy of a gene which is reinserted into the genome. Because of their origin, they do not contain any introns or promoters. It also lacks the nucleotide sequences immediately upstream of the 5′-UTR of the parent gene, which means that a processed pseudogene is inactive. The problem of discriminating between pseudogenes and frameshift sequencing errors is actually quite complex and will often be solved only by direct experimentation, e.g., gene expression studies or direct sequencing.

Among the large number of programs, algorithms, databases, and Web pages that can be used during the protein annotation process, there are some essential examples which are briefly detailed next:

NCBI. National Center for Biotechnology Information Web page (http://www.ncbi.nlm.nih.gov) is the largest source of information in molecular biology. Its databases are built from sequences submitted by individual researchers, genome sequencing consortiums, or data exchange with other databases. Among the databases maintained by NCBI are GenBank, PubMed, Online Mendelian Inheritance in Man (OMIM), the Molecular Modeling Database (MMDB) of 3D protein structures, the Unique Human Gene Sequence Collection (UniGene), a Gene Map of the Human Genome and the Taxonomy Browser, or the Cancer Genome Anatomy Project (CGAP). To search and retrieve sequences, NCBI uses the Entrez interface that provides access to sequence, mapping, taxonomy, and structural data *(33)*.

Ensembl (http://www.ensembl.org) is a joint project between the European Molecular Biology Laboratory (EMBL), the European Bioinformatics Institute (EBI), and the Wellcome Trust Sanger Institute (WTSI) to develop a software system which produces and maintains automatic analysis and annotation on selected eukaryotic genomes and the presentation of this kind of data via the Web. Ensembl also offers open software development *(34)*.

SMART (Simple Modular Architecture Research Tool http://smart.embl-heidelberg.de) allows the identification and annotation of protein domains and the analysis of domain architectures. Using either protein sequence or accession number, more than 500 extensively annotated domain families found in proteins, as well as signal peptides, transmembrane domains or low complexity segments are detectable *(35, 36)*.

PFAM (http://www.sanger.ac.uk/Software/Pfam) is a large collection of protein families, each represented by multiple sequence alignments and hidden Markov models (HMMs). It includes the choice of examining multiple alignments, protein domain architectures, and species distribution or known protein structures *(37)*.

MEROPS database (http://merops.sanger.ac.uk) is a specific information resource for proteases and their inhibitors. The MEROPS classification of proteases is done at the protein domain level and is hierarchical. Proteases that are statistically significant similar in amino acid sequences are grouped in a family, and families are grouped in a clan if there are indications that there was a common ancestor. The same principles have been used to classify protease inhibitors *(4, 38)*.

Degradome database (http://www.uniovi.es/degradome) comprises a set of manually and thoroughly curated protease and protease inhibitor coding genes from human, mouse, rat, and chimpanzee, including a catalog of human hereditary diseases of proteolysis or degradomopathies *(9, 39, 40)*.

PROSITE (http://www.expasy.ch/prosite) consists of documentation entries describing protein domains, families, and functional sites as well as associated patterns and profiles to identify them *(41)*.

InterPro (http://www.ebi.ac.uk/interpro) is a consortium of protein families' databases (PROSITE, Pfam, Prints, ProDom, SMART, and TIGRFAMs among others). Protein domains and functional sites found in annotated proteins can be predicted through computational methods in unknown protein sequence sites *(42)*.

As discussed above, similarity searches are performed to identify homologs of a given query protein or nucleotide sequence among all the sequences in a database. An alignment of homologous protein sequences reveals conserved regions important for their structure and function as well as poorly conserved. In principle, homologs are identified by aligning the query sequence against the database, sorting these hits based on similarity, and assessing the statistical significance that is likely to be indicative of homology. For this purpose, two general approaches can be initially considered.

The first option relies on comparison algorithms such as BLAST or BLAT to identify homologs or protein family members with a relatively high similitude with the protein used as query.

BLAST (Basic Local Alignment Search Tool) *(43)* is the most widely used method for sequence similarity searches in protein and DNA databases and statistical significance calculation of this similarity. Given two sequences, BLAST seeks for segments of a given length word size (W) within both sequences, which score at least T (threshold value) when they are aligned without gaps using a given substitution matrix. The expected value (E) is the number of matches that one should expect by chance. W, T, and E parameters, which can be changed by the user, determine the speed and sensitivity of the search. Then, word hits are extended in either direction in an attempt to generate a larger alignment. Usually, not only the perfect alignment is found, but also other locally optimal pairs, whose scores cannot be improved by extension. These locally optimal alignments are called "high-scoring segments pairs" or HSPs. So, instead of searching for perfect matches, BLAST looks for HSPs and statistically assesses if the hit has occurred by chance or is likely to be biologically relevant. The BLAST suite of programs is available either at the NCBI (http://www.ncbi.nlm.nih.gov/ BLAST), or at the Ensembl (http://www.ensembl.org/Multi /blastview) Web sites among other Web sites. BLAST software and databases can also be downloaded and locally run in any computer. This is particularly suitable for large-scale searches which are common in genome-wide analysis, because hundreds of queries can be automatically run or processed with simple scripts without bandwidth limitations. These BLAST programs can be used in any of the multiple available databases. Some examples are databases which contain complete genomic sequences from species, traces from genome projects, nonredundant protein sequences databases, protein sequences from Protein Data Bank or SwissProt, or Expressed Sequence Tags (ESTs) databases. Because of its speed and flexibility, BLAST is the first option when sequence similarity searches are required and it is the program most frequently used for genome analysis.

BLAT (BLAST Like Alignment Tool) *(44)* is a program specially intended for the nucleotide or protein alignment and comparison at a genomic scale. It is designed to rapidly and reliably find sequences with 95% and greater similarity of length 40 bases or more at nucleotide level and 80% and greater similarity of length 20 amino acids or more in proteins. BLAT and The UCSC Genome Browser *(45)* are available at the USCS Genome Bioinformatics Web page (http://genome.ucsc.edu), and the combination of both tools provides a rapid display of any region of included genomes, together with many annotation tracks like chromosome or contig positions, GC content, annotated genes, ESTs, SNPs, expression data, or conservation with other organisms.

Nevertheless, when the above methodologies fail to identify distantly related homologs because their sequences show a high divergence, a more sensitive approach is needed. This option takes advantage of the use of HMMs or position-specific score matrix (PSSMs).

A *Hidden Markov Model* (HMM) is a probabilistic model that captures position-specific information and can be applied to protein and DNA sequence pattern recognition. HMMs represent a system as a set of discrete states and as transitions between those states. Each transition has an associated probability. HMMs are valuable because they allow a search or alignment algorithm to be built on firm probabilistic bases, and its parameters (transition probabilities) can be easily trained on a known data set. HMMs can summarize the statistical properties of protein families and, through alignments for complete families, produce a consensus sequence that can be used to perform sensitive database searching. These HMM for protein families can be downloaded from PFAM Web page (http://www.sanger.ac.uk/Software/Pfam) or manually generated by HMMER software (http://hmmer.janelia.org) *(46)* from a multiple sequence alignment. Alignment methods are developed under the principle of hierarchical clustering that approximates the phylogenetic tree which guides the alignment. Once two sequences are aligned, compared, and clustered, they are treated as a single sequence. This step reduces the time and complexity in multiple sequences alignment processes. The most commonly used method for hierarchical multiple alignment is *Clustal (47)*, which is used in the ClustalW or ClustalX variants (available at http://www.ebi.ac.uk/clustal). Among the included programs in HMMER package, the most useful are *hmmbuild* that builds a model from a multiple sequence alignment; *hmmcalibrate* which takes an HMM and determines parameters to make searches more sensitive; and *hmmsearch* to search a sequence database for matches to an HMM.

PSI-BLAST (Position-Specific Iterated BLAST) combines alignments produced by BLAST with a PSSM This PSSM is automatically constructed from the multiple alignment obtained in an initial BLAST search, by calculating position-specific scores for every position. The more conserved position, the higher will be the score in the matrix. This PSSM is then used to run a second BLAST, with a new position-specific scores calculation step which successively refines the matrix. This iterative method results in a much more sensitive method to detect weak sequence relationships than BLAST *(48)*.

Taking this information into account, a general protocol for the annotation of protease coding genes based on computational approaches can be summarized as follows (**Fig. 1**):

1. To annotate the degradome of a new genome using homology-based comparisons, a set of reliably annotated proteases from another organism is needed. For this purpose, we built a non-redundant set of protease genes by combining information from literature, the MEROPS database, the proteome analysis database (http://www.ebi.ac.uk/proteome), and annotations derived from the experimental work at our laboratory.

2. Genomic sequences from a new nonannotated organism can be then analyzed for the presence of protease genes using TBLASTN and the nonredundant set of proteases to query the whole new genome. Every hit with *P* value under a pre-determined threshold (e.g., 10^{-2}) is analyzed by using the

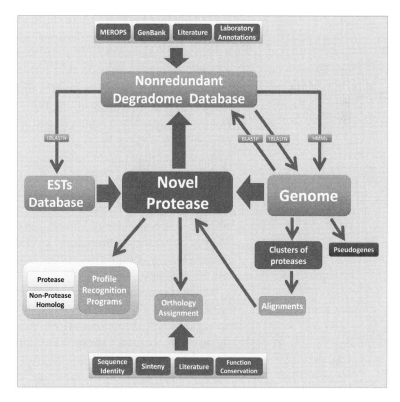

Fig. 1. Diagram of the general protocol for the annotation of the protease coding genes based on the use of computational algorithms and public databases.

BLASTP program against a custom degradome database that includes previously annotated proteases from other organisms. Hits not present in the custom database are further analyzed by using TBLASTN against the nonredundant nucleotide database at NCBI and TBLASTN of a 500,000-bp fragment containing the hit against similar proteases to build a predicted sequence. These strategies allow the extension of the putative protease fragments. Then, manual inspection against homologs and available EST sequences is used to complete the full ORF. Pseudogenes are defined by the presence of premature stop codons or frameshifts in sequences derived from the genome assembly, high-throughput genomic sequences at NCBI and EST sequences if available. Profile recognition programs, including SMART and InterPro can be used in this step to determine the presence of protease motifs, and multiple sequence alignments are used to finally classify a protein as protease or nonprotease homolog.

3. Orthology assignment is based on four different criteria: synteny (gene loci in different organisms that are located on a chromosomal region of common evolutionary ancestry); sequence identity based on reciprocal best match (genes A and B from genomes A′ and B′ are orthologs if B is the best match to A in genome A′ and A is the best match to B in genome B′), function conservation if possible, and relevant supporting literature. If one of these criteria is not met, a detailed analysis including conservation of neighbor genes in both species and examination of genome gaps, should be performed before orthology or paralogy assignment.

4. For each single hit, a 500-kb genomic sequence flanking the target gene should be analyzed for the presence of further members of that family, since approximately 23% of human, 34% of mouse, and 31% of rat protease genes are organized in clusters. A combination of detailed genomic sequence and relevant literature analysis should be used to determine the number of protease genes present in clusters of protease genes with high sequence similarity, as they can be artificially collapsed during genome assembly process. To establish orthology or paralogy in densely populated clusters of protease genes in which different members are specifically expanded, a phylogenetic tree is a helpful tool. Protein sequences corresponding to the full-length protease from different species are aligned by using the ClustalX program, together with a more distantly related protease to be used as root. Phylogenetic trees can be constructed for each family by using the Phylip package (http://evolution.genetics.washington.edu/phylip.html).

5. To further extend the bioinformatic search of protease genes, a HMM for the different protease families can be downloaded

from PFAM (http://www.sanger.ac.uk/Software/Pfam/) or manually built from proteases alignments that share a common family, using HMMER software. The selectivity of these models can be tested against the SWISS-PROT database, identifying known proteases when a *P*-value cutoff of 0.1 is used. Finally, analysis of the presence of ancillary domains in proteases is performed by using the SMART and Pfam domain databases.

3. Applications of Degradome Analysis for Mammalian Evolution Studies

The completed sequences of several mammalian genomes provide unprecedented choices to examine and characterize their degradomes (**Fig. 2**). Comparative genomics and bioinformatic approaches might lead to the identification of highly conserved elements or genetic differences that can be helpful in the elucidation of the molecular basis of diverse biological processes, the different susceptibility to diseases, or the evolution of the structure and function of proteases among mammals *(49)*.

3.1. The Human Degradome

A bioinformatic approach following the protocol described above was performed in the human genome to classify all previously known protease-coding genes and to identify novel

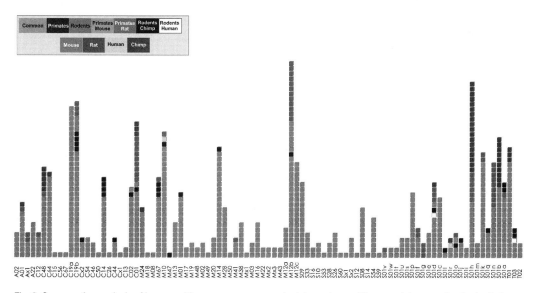

Fig. 2. Comparative analysis of human, chimpanzee, mouse, and rat degradomes. The complete nonredundant set of proteases and protease homologues from each species is distributed in five catalytic classes and 68 families. Each *square* represents a single protease and is *colored* according to its presence or absence in human, chimpanzee, mouse, and rat as indicated in the *inset. (See Color Plates)*

genes encoding proteins with sequence similarity to proteases from human or other organisms. This methodology allowed us to determine that the human degradome is composed of 569 proteases and protease-related genes grouped in 68 families, and 156 protease inhibitor genes, more than 2% of the total genes in the human genome, underscoring the importance of proteolysis in human biology. Human proteases can be divided into five different catalytic classes, with metalloproteases, serine and cysteine proteases being the most abundant (194, 176, and 150 genes in the human genome, respectively), while aspartic and threonine peptidases are composed of a limited number of members (21 and 28, respectively). Interestingly, among the 569 protease coding genes, 92 have lost key residues necessary for their proteolytic activity and have been classified as nonprotease homologs, and some of them might regulate the activation of other proteases or their access to substrates or inhibitors *(9)*.

3.2. Human vs. Rodents under a Proteolytic Prism

Due to the importance of model organisms, especially mouse (*Mus musculus*) and rat (*Rattus norvegicus*), in the knowledge of biological and pathological human processes, the availability of rodent genomes prompted us to identify their degradomes to gain insights into the evolution of mammalian proteases. The proteolytic repertoire analysis has provided the somewhat surprising result that rodent degradomes are more complex than the human degradome, despite their genomes are smaller *(9, 40)*. They are composed of 649 and 634 protease coding genes in mouse and rat, respectively, compared with the 569 proteases in human. As expected, the protease inhibitor repertoire has also undergone a marked expansion in rodents, with 199 genes in mouse and 183 in rat and only 156 in human. This higher complexity is essentially due to the expansion in rodents of specific families involved in reproduction and host defense, confirming that both processes have been important driving forces in mammalian evolution. The largest expansion has occurred in the kallikrein cluster (S01 serine proteases) that contains 15 protease genes in human, but 26 and 23 in mouse and rat. Similarly, the mast cell proteases family is constituted in humans by only 4 genes, whereas it has undergone an expansion in rodents with 17 genes in mouse and 28 in rat.

The higher number of protease genes found in rodent degradomes is also due to the specific inactivation of some protease coding genes in the human genome, while they are functional in mouse and rat. Thus, several proteases have been pseudogenized in humans, for example seven members of the ADAM family of metalloproteinases (Adam-1a, -1b, -3b, -4, -4b, -5, and 6) which are involved in ovum–sperm interaction, five testis serine proteases (Tessp3, Tessp6, and Tesp1, -2, and -3) and five digestive proteases (chymosin, Disp, trypsins Try10 and Try15, and pancreatic elastase Ela1). Probably, these pseudogenization processes are the result of a loss of function and subsequent accumulation of mutations *(9, 40)*.

3.3. Human vs. Chimpanzee Degradomes

The chimpanzee (*Pan troglodytes*) is our closest relative. Its degradome is composed of 559 protease coding genes, virtually identical to that of human. Despite the high conservation between proteases from both species, with 99.1% identities between human and chimpanzee orthologs, several proteases involved in immune functions show a higher degree of divergence, including neutrophil granule proteases, such as PRTN3 or GZMH, or proteases implicated in processing of proinflammatory cytokines such as CASP5. Furthermore, we have identified seven differential genes, most of them also related with host defense. Three of these genes are absent or have been pseudogenized in humans, but are functional in chimpanzees: NAPSB, CASP12, and HPP. The four remaining genes, GGTLA1, EOS, HPR, and MMP23A, are absent in the chimpanzee genome but are functional in humans. Despite these specific differences, we have also identified chimpanzee orthologs that are completely identical to their human counterparts. These absolutely conserved enzymes include proteases implicated in housekeeping processes, such as the proteasome components, or involved in neurological processes, such as PSEN1, BACE, and IMMP2L which are mutated in human diseases like Alzheimer's disease *(50, 51)* or Gilles de la Tourette syndrome *(52)*. These identities between proteases in both species suggest that the observed phenotypic differences might not be the result of changes in coding sequences, but changes in regulatory elements *(39, 53)*.

In summary, the degradome analysis of these organisms has allowed us to annotate their repertoire of protease coding genes, to identify genetic differences in mammalian species, and to assign them to specific biological processes. These analyses have shown that most of the differences are related to reproduction or host defense, suggesting that these two processes have been essential forces in the evolution of mammalian proteolytic systems. Despite these comparisons at the coding sequence level, to fully understand and explain biological differences in these organisms, we must take into account the importance of regulatory elements present in noncoding sequences, which lead to differences in gene expression and therefore in proteolysis between species. Thus, in addition to the annotation of protease coding genes in new sequenced genomes, future computational approaches in degradomics must include new tools for the study of regulatory elements in DNA sequences such as algorithms to detect transcription factor or miRNAs binding sites in protease promoters or 3′-UTRs, respectively, and therefore relate them to existing transcriptional data. Finally, in order to completely describe the proteolytic universe, the development of in silico methodologies aimed at the identification or prediction of protease substrates and inhibitors will facilitate the understanding of normal and pathological functions of proteases and the discovery of new therapeutic targets.

Acknowledgments

We thank Dr. J.P. Freije for helpful comments. Work in our laboratory is supported by grants from Ministerio de Educación y Ciencia, Fundación "La Caixa", Fundación Lilly, Fundación "M. Botín", and European Union (FP6 CancerDegradome).

References

1. Fujinaga, M., et al (2004) The molecular structure and catalytic mechanism of a novel carboxyl peptidase from *Scytalidium lignicolum*. *Proc Natl Acad Sci USA* 101(10), 3364–9.

2. Gomis-Ruth, F.X. (2003) Structural aspects of the metzincin clan of metalloendopeptidases. *Mol Biotechnol* 24(2), 157–202.

3. Polgar, L. (2005) The catalytic triad of serine peptidases. *Cell Mol Life Sci* 62(19–20), 2161–72.

4. Rawlings, N.D., D.P. Tolle, and A.J. Barrett (2004) MEROPS: the peptidase database. *Nucleic Acids Res* 32(Database issue), D160–4.

5. Barrett, A.J., N.D. Rawlings, and J.F. Woessner (2004) *Handbook of proteolytic enzymes, 2nd ed.* Amsterdam: Elsevier.

6. Hooper, N.M. (2002) Proteases in biology and medicine. London: Portland Press.

7. López-Otín, C. and L.M. Matrisian (2007) Emerging roles of proteases in tumour suppression. *Nat Rev Cancer* 7(10), 800–8.

8. Overall, C.M. and C. López-Otín (2002) Strategies for MMP inhibition in cancer: innovations for the post-trial era. *Nat Rev Cancer* 2(9), 657–72.

9. Puente, X.S., et al. (2003) Human and mouse proteases: a comparative genomic approach. *Nat Rev Genet* 4(7), 544–58.

10. Bowen, D.J. (2002) Haemophilia A and haemophilia B: molecular insights. *Mol Pathol* 55(2), 127–44.

11. Sherrington, R., et al. (1995) Cloning of a gene bearing missense mutations in early-onset familial Alzheimer's disease. *Nature* 375(6534), 754–60.

12. Whitcomb, D.C., et al. (1996), A gene for hereditary pancreatitis maps to chromosome 7q35. *Gastroenterology* 110(6), 1975–80.

13. Varela, I., et al. (2005) Accelerated ageing in mice deficient in Zmpste24 protease is linked to p53 signalling activation. *Nature* 437(7058), 564–8.

14. López-Otín, C. and C.M. Overall (2002) Protease degradomics: a new challenge for proteomics. *Nat Rev Mol Cell Biol* 3(7), 509–19.

15. TCSAC. (2005) Initial sequence of the chimpanzee genome and comparison with the human genome. *Nature* 437(7055), 69–87.

16. Gibbs, R.A., et al. (2004) Genome sequence of the Brown Norway rat yields insights into mammalian evolution. *Nature* 428(6982), 493–521.

17. Lander, E.S., et al. (2001), Initial sequencing and analysis of the human genome. *Nature* 409(6822), 860–921.

18. Venter, J.C., et al. (2001) The sequence of the human genome. *Science* 291(5507), 1304–51.

19. Waterston, R.H., et al. (2002) Initial sequencing and comparative analysis of the mouse genome. *Nature* 420(6915), 520–62.

20. Lewis, S., M. Ashburner, and M.G. Reese (2000) Annotating eukaryote genomes. *Curr Opin Struct Biol* 10(3), 349–54.

21. Stein, L. (2001) Genome annotation: from sequence to biology. *Nat Rev Genet* 2(7), 493–503.

22. Reed, J.L., et al. (2006) Towards multidimensional genome annotation. *Nat Rev Genet* 7(2), 130–41.

23. Brent, M.R. (2005) Genome annotation past, present, and future: how to define an ORF at each locus. *Genome Res* 15(12), 1777–86.

24. Shmatkov, A.M., et al. (1999) Finding prokaryotic genes by the 'frame-by-frame' algorithm: targeting gene starts and overlapping genes. *Bioinformatics* 15(11), 874–86.

25. Claverie, J.M. (1997) Computational methods for the identification of genes in vertebrate genomic sequences. *Hum Mol Genet* 6(10), 1735–44.

26. Burge, C.B. and S. Karlin (1998) Finding the genes in genomic DNA. *Curr Opin Struct Biol* 8(3), 346–54.

27. Zhang, M.Q. (2002) Computational prediction of eukaryotic protein-coding genes. *Nat Rev Genet* 3(9), 698–709.

28. Fitch, W.M. (2000) Homology a personal view on some of the problems. *Trends Genet* 16(5), 227–31.

29. Lynch, M. and J.S. Conery (2000) The evolutionary fate and consequences of duplicate genes. *Science* 290(5494), 1151–5.

30. Wagner, A. (2002) Selection and gene duplication: a view from the genome. *Genome Biol* 3(5), 1012.

31. Altschul, S.F. (1991) Amino acid substitution matrices from an information theoretic perspective. *J Mol Biol* 219(3), 555–65.

32. Henikoff, S. and J.G. Henikoff (1992) Amino acid substitution matrices from protein blocks. *Proc Natl Acad Sci USA* 89(22), 10915–9.

33. Wheeler, D.L., et al. (2007) Database resources of the National Center for Biotechnology Information. *Nucleic Acids Res* 35(Database issue), D5–12.

34. Hubbard, T.J., et al. (2007) Ensembl 2007. *Nucleic Acids Res* 35(Database issue), D610–7.

35. Letunic, I., et al. (2006) SMART 5: domains in the context of genomes and networks. *Nucleic Acids Res* 34(Database issue), D257–60.

36. Schultz, J., et al. (1998) SMART, a simple modular architecture research tool: identification of signaling domains. *Proc Natl Acad Sci USA* 95(11), 5857–64.

37. Finn, R.D., et al. (2006) Pfam: clans, web tools and services. *Nucleic Acids Res* 34(Database issue), D247–51.

38. Rawlings, N.D., F.R. Morton, and A.J. Barrett (2006) MEROPS: the peptidase database. *Nucleic Acids Res* 34(Database issue), D270–2.

39. Puente, X.S., et al. (2005) Comparative genomic analysis of human and chimpanzee proteases. *Genomics* 86(6), 638–47.

40. Puente, X.S. and C. López-Otín (2004) A genomic analysis of rat proteases and protease inhibitors. *Genome Res* 14(4), 609–22.

41. Hulo, N., et al. (2006) The PROSITE database. *Nucleic Acids Res* 34(Database issue), D227–30.

42. Mulder, N.J., et al. (2007) New developments in the InterPro database. *Nucleic Acids Res* 35(Database issue), D224–8.

43. Altschul, S.F., et al. (1990) Basic local alignment search tool. *J Mol Biol* 215(3), 403–10.

44. Kent, W.J. (2002) BLAT – the BLAST-like alignment tool. *Genome Res* 12(4), 656–64.

45. Kent, W.J., et al. (2002) The human genome browser at UCSC. *Genome Res* 12(6), 996–1006.

46. Eddy, S.R. (1998) Profile hidden Markov models. *Bioinformatics* 14(9), 755–63.

47. Thompson, J.D., D.G. Higgins, and T.J. Gibson (1994) CLUSTAL W: improving the sensitivity of progressive multiple sequence alignment through sequence weighting, position-specific gap penalties and weight matrix choice. *Nucleic Acids Res* 22(22), 4673–80.

48. Altschul, S.F., et al. (1997) Gapped BLAST and PSI-BLAST: a new generation of protein database search programs. *Nucleic Acids Res* 25(17), 3389–402.

49. Puente, X.S., et al. (2005) A genomic view of the complexity of mammalian proteolytic systems. *Biochem Soc Trans* 33(Pt 2), 331–4.

50. Esler, W.P. and M.S. Wolfe (2001) A portrait of Alzheimer secretases – new features and familiar faces. *Science* 293(5534), 1449–54.

51. Vassar, R. (2002), Beta-secretase (BACE) as a drug target for Alzheimer's disease. *Adv Drug Deliv Rev* 54(12), 1589–602.

52. Petek, E., et al. (2001) Disruption of a novel gene (IMMP2L) by a breakpoint in 7q31 associated with Tourette syndrome. *Am J Hum Genet* 68(4), 848–58.

53. Puente, X.S., et al. (2006) Comparative analysis of cancer genes in the human and chimpanzee genomes. *BMC Genomics* 7(1), 15.

Chapter 3

Microarrays for Protease Detection in Tissues and Cells

Kamiar Moin, Donald Schwartz, Stefanie R. Mullins, and Bonnie F. Sloane

Summary

Expression of a given protease and of the endogenous inhibitors that regulate protease activity can be readily determined at the transcript level by using whole genome microarray chips. In the case of proteases and protease inhibitors, however, determining which cells are expressing them is often critical to understanding the functional roles of the proteases. For example, in cancer many of the proteases are derived from cells that are found in the microenvironment surrounding the tumor, e.g., fibroblasts and inflammatory cells. Proteases from both fibroblasts and inflammatory cells have been implicated in malignant progression. Therefore, it is important to recognize the origin of these molecules if one is to develop effective therapies. In this regard, mouse transgenic models and xenograft models in which human tumor cells are implanted in mice are useful tools. To profile human and mouse proteases, protease inhibitors, and protease interactors, we have developed in partnership with Affymetrix a custom, single platform, dual species chip: the Hu/Mu ProtIn chip. The Hu/Mu ProtIn chip has been validated for its ability to identify human and mouse transcripts in single species specimens and to identify and distinguish between human and mouse transcripts in dual species specimens such as xenografts. In the latter specimens, the Hu/Mu ProtIn chip has enabled us to identify host (mouse) proteases that play a protective role in development of lung tumors. Here we outline a protocol for using the Hu/Mu ProtIn chip to profile proteases, protease inhibitors, and protease interactors in tissues and cells.

Key words: Protease, Protease inhibitors, Protease interactors, Microarray, Hu/Mu ProtIn chip, Xenograft, Tumor microenvironment.

1. Introduction

Proteases are involved in the regulation of physiologic and pathologic processes including inflammation, angiogenesis, tumorigenesis, invasion, and metastasis (reviewed in **refs.** *1–5*). On the basis of their common evolutionary origin, proteases have been divided into 47 clans of more than 175 families that share similar tertiary structures or catalytic sites *(6)*. Detailed information on known proteases and their relationships can be found in the MEROPS

Toni M. Antalis and Thomas H. Bugge (eds.), *Methods in Molecular Biology, Proteases and Cancer, Vol. 539*
© Humana Press, a part of Springer Science + Business Media, LLC 2009
DOI: 10.1007/978-1-60327-003-8_3

database online (http://merops.sanger.ac.uk,*6*). The activities of proteases are regulated by various endogenous protease inhibitors (for review *see* **refs.** *7–9).* Like their protease counterparts, the inhibitors are numerous and diverse, and have been classified into 33 clans of 55 families *(6).* Other proteins, such as activators and receptors, affect the biological functions of proteases by modulating their behavior, trafficking, and localization.

To understand the significance of proteolysis in any physiologic and/or pathologic process, one has to consider proteolysis in the context of the tissue microenvironment and the interplay therein among proteases, protease inhibitors, and protease interactors. Suitable model systems that can be experimentally manipulated are helpful as they provide the ability to compare proteolytic pathways under various conditions. One example for elucidating tumor–host interactions is xenografts of human cancer cells implanted orthotopically in mice. Furthermore, to determine the role played by proteases, protease inhibitors, and protease interactors in any given process, whether pathologic or physiologic, they first need to be identified and their levels of expression in the affected tissues/cells measured.

Oligonucleotide microarrays have greatly facilitated our ability to measure levels of gene expression in diseases such as cancer (for review, *see* **ref.** *10).* The increasing evidence that development and progression of cancers is affected both positively and negatively by interactions between the tumor and its microenvironment *(11–16)* suggests that the ability to profile gene expression in both compartments will be critical. Xenograft models of human cancer provide a system in which the genes in the two compartments are from different species: host (mouse) and graft (human). Typical oligonucleotide microarrays contain probes for profiling gene expression from only one species and as a consequence expression of both host and graft genes cannot be determined simultaneously (*see* **Note 1**). Although one could profile sequentially with two microarrays, one specific for human and one for mouse, this doubles the expense. More importantly, such dual-species analyses are likely to lead to misinterpretation as probes on single-species arrays are not designed to prevent binding of transcripts from other species but rather to profile transcripts of only one species. Therefore, cross-hybridization could be interpreted as a change in expression (*see* **Note 1**). To address this, in partnership with Affymetrix, we have designed and developed a dual species oligonucleotide microarray, the Hu/Mu ProtIn chip, to profile human and murine protease, protease inhibitor, and protease interactor genes. All of the probes on the custom Hu/Mu ProtIn chip were blasted against both human and mouse genome sequences to minimize cross-species hybridization *(17).* The ability of the Hu/Mu ProtIn chip to differentiate murine from human and human from murine gene expression and the specificity and sensitivity of the probe sets on this chip was determined based on comparative gene expression analyses in pure mouse or human samples and orthotopic xenograft models of human breast and lung cancers (**Fig. 1**, *(17)*). Thus, the Hu/Mu ProtIn chip, although designed for interrogation of xenograft samples, also can determine expression of proteases, protease inhibitors,

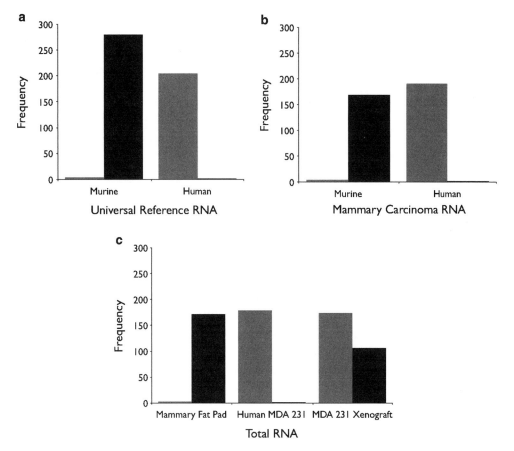

Fig. 1. Validation of the Hu/Mu ProtIn chip: (**a**) Three independent replicates of XpressRef (SuperArray Bioscience) universal reference total RNA from mouse or human tissues were profiled with the Hu/Mu ProtIn microarray. The presence of a transcript was determined by the detection algorithm in MAS5 software (Affymetrix), which assigns P as a measure of the detection call confidence. The number of probe sets, giving present call *P* values smaller than 0.01, for all three replicates are indicated. In the presence of only mouse transcripts, there were extremely few present detection calls by human probe sets (*gray bars*), and vice versa (*black bars*), despite the high sequence similarity among protease and protease inhibitor genes and their potential for cross-hybridization. (**b**) Total RNA derived from spontaneous mammary carcinomas from MMTV-PyMT⁺ (FVB/n) (*n* = 8) and MMTV-PyMT⁺/uPARAP⁻/⁻ (FVB/n) (*n* = 2) transgenic mice or human breast ductal carcinoma biopsies (*n* = 18) from women with stage II or III disease was profiled. The number of probe sets, giving present call *P* values smaller than 0.01, for ≥80% are indicated. Transcripts derived from mammary carcinoma were rarely detected by nonspecies identical probes (human, *gray bars*; mouse, *black bars*). In C, total RNA derived from normal mouse mammary fat pads (*n* = 4), MDA-MB-231 breast carcinoma cells (*n* = 3), and orthotopically implanted xenografts of MDA-MB-231 cells (*n* = 3) was profiled. Number of probe sets, giving present call *P* values smaller than 0.01, for all replicates in each group is indicated. Nonspecies-identical probes (*gray bars*, human and *black bars*, mouse) rarely detect transcripts in mouse mammary fat pads, and MDA-MB-231 cells. Importantly, transcripts from MDA-MB-231 orthotopic xenografts were detected by both mouse (*black bars*) and human probes (*gray bars*), suggesting that both mouse (host) and human (tumor) transcripts present in the xenograft can be detected simultaneously by the Hu/Mu ProtIn microarray. Frequency represents the number of probe sets that satisfy the selection criterion. Adapted from **ref.***17*

and protease interactors in single-species human and mouse samples (*see* **Note 1**). Furthermore, the Hu/Mu ProtIn chip has successfully identified proteases that function in proteolytic processes that protect against development of murine tumors of the intestine *(18)* and the lung *(19)*. The former study was a single species analysis of mouse tissues, and the latter was a dual-species analysis of human tumor cells orthotopically implanted in mouse lung. In both, the Hu/Mu ProtIn chip proved an important tool for the identification of new tumor suppressive roles for proteases in cancer *(20)*, i.e., roles that in the case of matrix metalloproteinases may explain the failure of the clinical trials using broad spectrum inhibitors of these enzymes *(4)*.

Here we outline protocols for using the Hu/Mu ProtIn chip to profile proteases, protease inhibitors, and protease interactors in tissues and cells.

2. Materials

2.1. Sample Preparation and RNA Isolation

1. Tissue, fresh or snap-frozen.
2. Cultured cells grown as monolayers or in a 3D matrix.
3. RNALater reagent (Ambion, Austin, TX).
4. QIAshredder tissue homogenizer (Qiagen, Valencia, CA) or equivalent.
5. Dulbecco's phosphate buffered saline (PBS) without calcium or magnesium (Invitrogen, Carlsbad, CA).
6. Dispase (BD Biosciences, San Jose, CA).
7. Collagenase (Invitrogen, Carlsbad, CA).
8. TRIZOL™ (Invitrogen, Carlsbad, CA).
9. RNeasy Mini-Kits (Qiagen, Valencia, CA).
10. RNAse free DNAse set (Qiagen, Vaencia, CA).
11. Hu/Mu ProtIn chip: Protease520066F (Affymetrix, Santa Clara, CA).
12. XpressRef universal reference total RNA from mouse or human tissues (SuperArray Bioscience, Frederick, MD).

3. Methods

In general, microarray chips are processed and analyzed by an institutional genomics core facility. This is in part due to the fact that high-end, specialized instrumentation is needed that is beyond the research budget of most individual investigators. Core facilities may prefer to handle the entire protocol (from RNA isolation through chip processing) to ensure adequate quality control.

3.1. Sample Preparation and RNA Isolation

Total RNA can be isolated in one's laboratory or the samples may be given to a genomics core for isolation of RNA. If RNA is to be prepared in the laboratory, one must meet the minimum requirements of the genomics core in regard to RNA integrity and purity. Tissue samples or cultured cells may be fresh or frozen (snap frozen and stored at –80°C) as long as ample precaution has been exercised to protect RNA from degradation. The following are the protocols we have utilized successfully in our laboratory to provide the highest quality RNA to our genomics facility.

3.1.1. Isolation of RNA from Tissue Samples

1. Place all tissue samples, fresh or frozen, immediately in RNALater and store for at least 24 h at 4°C. To ensure rapid and sufficient diffusion of RNALater throughout the tissue, make sure that no tissue dimension exceeds 0.5 cm.

2. Centrifuge briefly at 300 g (2 min) to remove the preservative and freeze the sample at –80°C. Transfer tissue (50–100 mg) to a QIAshredder tissue homogenizer or equivalent and briefly centrifuge (1,000 × g, 2 min) at room temperature to remove residual RNALater.

3. Remove the tissue from the QIAshredder and transfer to a nuclease-free microcentrifuge tube containing 1 mL of TRIZOL™ on ice.

4. Isolate total RNA from TRIZOL™ as recommended by the manufacturer.

5. Following isolation, further purify RNA with an RNeasy Mini-Kit.

3.1.2. Isolation of RNA from Monolayer Cell Cultures

1. Discard the media.

2. Wash the cells with Dulbecco's PBS for three times.

3. Aspirate the residual wash.

4. Remove the cells directly into 1 mL TRIZOL™ per 35-cm² dish, using a plastic cell scraper to aid in lysing the cells.

5. Collect cell lysates and incubate at room temperature for 10 min.

6. Store the lysates at –80°C until RNA isolation.

7. Follow **steps 4** and **5** in **Subheading 3.1.1** to isolate RNA.

3.1.3. Isolation of RNA from 3D Cell Cultures

If cells are grown in a 3D extracellular matrix such as Matrigel (BD Bioscience, San Jose, CA), collagen I (BD Bioscience, San Jose, CA) or Cultrex (Trevigen, Gaithersburg, MD), the following procedure should be followed to remove the matrix.

1. Incubate 3D cultures with 1 U Dispase or 0.1% w/v collagenase (if the matrix is collagen) in Dulbecco's PBS at 37°C for approximately 20 min to release the cells.

2. Centrifuge briefly (300 × g, 2 min) and wash the cells three times with Dulbecco's PBS.

3. Resuspend cell pellet in 1 mL TRIZOL™.

4. Follow **steps 5–7** in **Subheading 3.1.2** above to isolate RNA.

3.2. RNA Integrity and Yield

At this point the purity, integrity, and quantity of RNA needs to be determined by the genomics core facility. Please note that in most cases this is a requirement set by the facility because of the high cost of microarray processing and analysis. In general, any sample with an rRNA ratio of <1 (as determined by the core) is not suitable for processing.

3.3. RNA Processing, cRNA Production, and cRNA Labeling

Since the Hu/Mu ProtIn chip is based on an Affymetrix platform, the processing of RNA for hybridization, including cRNA synthesis and labeling, has to precisely follow standard protocols described in the Affymetrix GeneChip Expression Analysis Technical Manual. These are performed with platform-specific instrumentation available only through genomic cores.

3.4. Data Analysis

Analysis of the expression data obtained from the Hu/Mu ProtIn chip or any other microarray chip requires extensive bioinformatics expertise and sophisticated computer software. Again, as with chip processing, unless there is bioinformatics expertise in one's laboratory, data should be transmitted to a bioinformatics core facility for analysis. We have used Microarray Suite version 5 (MAS5, Affymetrix) *(21)*, dChip *(22)*, and trimmed-mean algorithm (TM) methods *(23, 24)*, to characterize and validate the Hu/Mu ProtIn chip. Details of this extensive analysis are reported in **ref.** *17*.

3.4.1. Calculations

1. In MAS5, set expression analysis parameters to default values (e.g., scale factor = 500, normalization = 1.0 and no masking).

2. Generate a p value for the likelihood of the presence of a transcript. The detection algorithm in MAS5 software utilizes a one-sided Wilcoxon's signed rank test of the discrimination value (R) relative to a user-defined threshold Tau (default = 0.015).

3. Calculate the Discrimination value by adjusting the intensity difference, PM (Perfect Match) – MM (Mismatch), to the overall hybridization intensity (PM + MM). A Discrimination value greater than Tau indicates the presence of the transcript and less than Tau for the absence of the transcript.

4. In dChip, calculate the frequency of a detection call. Make sure that 100% of the samples are present at the p value generated in **Step 2** (e.g., 0.05).

3.5. Validation of Probe Design

To compare arrays of different samples, the raw data from each array need to be normalized using the invariant set normalization algorithm within the dChip software. Gene expression values are calculated using the Model Based Expression Index (MBEI). For some experiments, normalization and MBEI can be performed with PM and MM probe information (dChip PM + MM). In some experiments, gene expression values may be calculated using a TM followed by quantile normalization to a standard array as previously reported *(18)*. TM can be performed by removing either 20 (TM20) or 25 (TM25) percent of the PM-MM probe pair values within each probe set prior to calculating the gene expression value *(17)*.

3.6. Verification of Expression

Although the Hu/Mu protIn chip has been extensively characterized and validated *(17)*, any microarray expression data should be verified through other techniques. At the transcript level, the data can be validated with RT-PCR utilizing TaqMan primers *(25)*. Ultimately, however, validation must be accomplished at the protein level through the use of tissue microarrays or tissue sections through immunohistochemical analysis.

4. Notes

1. **Alternative methods.** Proteases, protease inhibitors, and protease interactors can be profiled in human samples, mouse samples, and samples containing both human and mouse genes by the sequential use of human genome chips and mouse genome chips available commercially. These single-species chips, however, are designed to profile transcripts of only one species and may misinterpret cross-hybridization as a change in expression. Costs will of course be double for analysis with two chips. Proteases and protease inhibitors, but not protease interactors, can be profiled in human samples, mouse samples, and samples containing both human and mouse genes with the hybrid version of the Clip-Chip™, a spotted oligonucleotide array, which is available through collaboration with Overall and colleagues *(26, 27)*. One advantage of this chip is that it contains probe sets for more proteases and inhibitors than on the Hu/Mu ProtIn chip. At this time, however, the hybrid Clip-Chip™ "is still under validation to determine cross-species hybridization between the oligonucleotides" *(27)*. Thus, whether the Clip-Chip™

will prove suitable for analysis of xenografts is not yet known. In a dual-species array when proteases and protease inhibitors of the two species are known to be highly homologous *(7, 28, 29)*, the use of shorter probes, e.g., the 25-mer probes used for the Hu/Mu ProtIn chip, is desirable as longer probes increase the probability for cross-hybridization. Sensitivity is higher with longer probes such as the 70-mer probes used in the Clip-Chip™. When 25- and 60-mer probes were compared, the 25-mer probes were found to be 20-fold more specific and the 60-mer probes threefold more sensitive *(30)*, indicating that the Clip-Chip™ will be less likely to distinguish expression of human and mouse genes. Depending on the design of individual experiments, either the Hu/Mu ProtIn or Clip-Chip™ array may be most appropriate or one may want to use the two arrays in tandem.

Acknowledgment

This work was supported by a Department of Defense Breast Cancer Center of Excellence (DAMD17–02–1–0693).

References

1. DeClerck YA, Mercurio AM, Stack MS, Chapman HA, Zutter M M, Muschel RJ, Raz A, Matrisian LM, Sloane BF, Noel A, Hendrix MJ, Coussens L, and Padarathsingh M. (2004) Proteases, extracellular matrix, and cancer: a workshop of the path B study section. *Am J Pathol* 164, 1131–1139.

2. Lynch CC and Matrisian LM. (2002) Matrix metalloproteinases in tumor-host cell communication. *Differentiation* 70, 561–573.

3. van Kempen LC, de Visser KE, and Coussens LM. (2006) Inflammation, proteases and cancer. *Eur J Cancer* 42, 728–734.

4. Coussens LM, Fingleton B, and Matrisian LM. (2002) Matrix metalloproteinase inhibitors and cancer: trials and tribulations. *Science* 295, 2387–2392.

5. Podgorski I and Sloane BF. (2003) Cathepsin B and its role(s) in cancer progression. *Biochem Soc Symp* 70, 263–276.

6. Rawlings ND, Morton FR, Kok CY and Barrett AJ. (2007) MEROPS: the peptidase database. *Nucleic Acids Res* 35, 1–6.

7. Abrahamson M, Alvarez-Fernandez M, and Nathanson CM. (2003) Cystatins. *Biochem Soc Symp* 70, 179–199.

8. van Gent D, Sharp P, Morgan K, and Kalsheker N. (2003) Serpins: structure, function and molecular evolution. *Int J Biochem Cell Biol* 35, 1536–1547.

9. Murphy G, Knauper V, Lee MH, Amour A, Worley JR., Hutton M, Atkinson S, Rapti, M, and Williamson, R. (2003) Role of TIMPs (tissue inhibitors of metalloproteinases) in pericellular proteolysis: the specificity is in the detail. *Biochem Soc Symp* 70, 65–80.

10. Bucca G, Carruba G, Saetta A, Muti P, Castagnetta L, and Smith CP. (2004) Gene expression profiling of human cancers. *Ann N Y Acad Sci* 1028, 28–37.

11. Tlsty TD and Coussens LM. (2006) Tumor stroma and regulation of cancer development. *Annu Rev Pathol* 1, 119–150.

12. Radisky ES, Radisky DC. (2007) Stromal induction of breast cancer: inflammation and invasion. *Rev Endocr Metab Disord* 8, 279–287.

13. Egeblad M, Littlepage LE, Werb Z. (2005) The fibroblastic coconspirator in cancer progression. *Cold Spring Harb Symp Quant Biol* 70, 383–388.

14. Jodele S, Blavier L, Yoon JM, DeClerck YA. (2006) Modifying the soil to affect the seed: role of stromal-derived matrix metalloproteinases in cancer progression. *Cancer Metastasis Rev* 25, 35–43.

15. Sloane BF, Yan S, Podgorski I, Linebaugh BE, Cher ML, Mai J, Cavallo-Medved D, Sameni M, Dosescu J, Moin K. (2005) Cathepsin B and tumor proteolysis: contribution of the tumor microenvironment. *Semin Cancer Biol* 15, 149–157.

16. Overall CM, Kleifeld O. (2006) Tumour microenvironment - opinion: validating matrix metalloproteinases as drug targets and antitargets for cancer therapy. *Nat Rev Cancer* 6, 227–39.

17. Schwartz DR, Moin K, Yao B, Matrisian LM, Coussens LM, Bugge TH, Fingleton B, Acuff HB, Sinnamon M, Nassar H, Platts AE, Krawetz SA, Linebaugh BE, Sloane BF. (2007) Hu/Mu ProtIn oligonucleotide microarray: dual-species array for profiling protease and protease inhibitor gene expression in tumors and their microenvironment. *Mol Cancer Res* 5, 443–454.

18. Sinnamon MJ, Carter KJ, Sims LP, LaFleur B, Fingleton B, and Matrisian LM. (2007) A protective role for mast cells mediated through eosinophils in intestinal tumorigenesis. *Carcinogenesis* 4, 880–886.

19. Acuff HB, Sinnamon M, Fingleton B, Boone B, Levy SE, Chen X, Pozzi A, Carbone DP, Schwartz DR, Moin K, Sloane BF, Matrisian LM. (2006) Analysis of host- and tumor-derived proteinases using a custom dual species microarray reveals a protective role for stromal matrix metalloproteinase-12 in non-small cell lung cancer. *Cancer Res* 66, 7968–7975.

20. López-Otín C, Matrisian LM. (2007) Emerging roles of proteases in tumour suppression. *Nat Rev Cancer* 7, 800–808.

21. Hubbell E, Liu WM, and Mei R. (2002) Robust estimators for expression analysis. *Bioinformatics* 18, 1585–1592.

22. Li C and Wong WH. (2001) Model-based analysis of oligonucleotide arrays: expression index computation and outlier detection. *Proc Natl Acad Sci USA* 98, 31–36.

23. Giordano TJ., Shedden KA, Schwartz DR, Kuick R, Taylor JM, Lee N, Misek DE, Greenson JK, Kardia SL, Beer DG, Rennert G, Cho KR, Gruber SB, Fearon ER, and Hanash S. (2001) Organ-specific molecular classification of primary lung, colon, and ovarian adenocarcinomas using gene expression profiles. *Am J Pathol* 159, 1231–1238.

24. Rickman DS, Bobek MP, Misek DE, Kuick R, Blaivas M, Kurnit DM, Taylor J, and Hanash SM. (2001) Distinctive molecular profiles of high-grade and low-grade gliomas based on oligonucleotide microarray analysis. *Cancer Res* 61, 6885–6891.

25. Pennington C, Nuttall R, Sampieri C, Wallard M, Pilgrim S and Edwards DR. Quantitative Real-Time PCR analysis of degradome gene expression. In: *The Cancer Degradome—Proteases and Cancer Biology*. DR Edwards, GF Hoyer-Hansen, F Blansi and BF Sloane eds., Springer, New York, 2008.

26. Overall CM, Tam EM, Kappelhoff R, Connor A, Ewart T, Morrison CJ, Puente X, Lopez-Otin C, and Seth A. (2004) Protease degradomics: mass spectrometry discovery of protease substrates and the CLIP-CHIP, a dedicated DNA microarray of all human proteases and inhibitors. *Biol Chem* 385, 493–504.

27. Kappelhoff R, Wilson C, Overall CM. The CLIP-CHIP™. In: *The Canacer Degradome—Proteases and Cancer Biology*. DR Edwards, GF Hoyer-Hansen, F Blansi and BF Sloane eds., Springer, New York, 2008.

28. Nagase H, Brew K. (2003) Designing TIMP (tissue inhibitor of metalloproteinases) variants that are selective metalloproteinase inhibitors. *Biochem Soc Symp* 70, 201–212.

29. Puente XS, Sanchez LM, Overall CM and López-Otín C. (2003) Human and mouse proteases: a comparative genomic approach. *Nature Rev Genetics* 4, 544–558.

30. Relogio A, Schwager C, Richter A, Ansorge W, Valcárcel J. (2002) Optimization of oligonucleotide-based DNA microarrays. *Nucleic Acids Res* 30, e51.

31. Schwartz DR, Kardia SL, Shedden KA, Kuick R, Michailidis G, Taylor JM, Misek DE, Wu R, Zhai Y, Darrah DM, Reed H, Ellenson LH, Giordano TJ, Fearon ER, Hanash SM, and Cho KR. (2002) Gene expression in ovarian cancer reflects both morphology and biological behavior, distinguishing clear cell from other poor-prognosis ovarian carcinomas. *Cancer Res* 62, 4722–4729.

Chapter 4

Positional Scanning Synthetic Combinatorial Libraries for Substrate Profiling

Eric L. Schneider and Charles S. Craik

Summary

Determining the preferred substrate cleavage sequence of proteases is an important step toward under-standing their roles in cancer development and progression. Knowledge of this sequence can aid in the design of new experimental tools for study as well as aid in the identification of endogenous protease substrates and signaling pathways. Various investigators have demonstrated a number of techniques to uncover these sequences, but most can be very time consuming. We have designed and successfully implemented a complete diverse ACC tetrapeptide positional scanning synthetic combinatorial library that allows for the rapid screening of proteases to determine their preferred residues at positions P1-P4. These sequences can be readily verified through kinetic measurements on single peptide substrates and utilized to further knowledge of the role of proteases in cancer.

Key words: Peptide library, Substrate profiling, Protease specificity, PS-SCL, ACC.

1. Introduction

Understanding the *in vivo* role and function of proteases has become a promising avenue of research for identifying diagnostics, prognostics, and therapeutic targets in the fight against cancer. Although there have been a number of proteases identified in the various stages of cancer growth and progression, determining the precise role that each one plays has proven to be very challenging. One way to provide insight regarding these aspects is to identify the endogenous substrate(s) of each protease. Biochemical and proteomic methods can be used to identify a preferred cleavage sequence specific to a protease, which can in turn be used to identify the endogenous substrates and signaling pathways that

Toni M. Antalis and Thomas H. Bugge (eds.), *Methods in Molecular Biology, Proteases and Cancer, Vol. 539*
© Humana Press, a part of Springer Science + Business Media, LLC 2009
DOI: 10.1007/978-1-60327-003-8_4

involve the protease of interest. Unfortunately, many of the traditional methods for this can be time consuming. Positional scanning synthetic combinatorial libraries (PS-SCLs) of fluorogenic peptides offer the promise to expedite the initial steps toward identification of these substrates.

Positional scanning synthetic combinatorial libraries contain a mixture of peptides where one or more of the positions are individually fixed at a specific amino acid, while the remaining positions are composed of an equimolar mixture of amino acids. Through the use of combinatorial chemistry, these libraries can be quickly synthesized and analyzed. Dependent upon the number of randomized positions (X) and number of amino acids incorporated in those positions (Y), there will be Y^X different peptides in each mixture, with each containing the same fixed amino acid(s). The benefit of this preparation is the ability to determine the effect of the fixed amino acids independent of the residues in the remaining positions. Initial studies preparing and utilizing these libraries efficiently screened over 34 million hexapeptides to identify the antigenic determinant of a monoclonal antibody *(1)*. This study demonstrated the utility of these libraries to rapidly screen through large numbers of peptides and determine the optimal sequences for peptide binding.

Because the PS-SCL approach proved to be very useful in scanning through large numbers of peptides, fluorogenic leaving groups were added to screen against proteases. The original libraries were created with an aminomethylcoumarin (AMC) fluorogenic leaving group attached to the C-terminus (the P1 amino acid) of the peptide via an amide bond *(2, 3)*. Cleavage of the amide bond results in increased fluorescence of the free AMC group, providing an ideal reporter for enzyme activity. The initial libraries were successfully used to identify the substrate specificities of caspases 1–9, granzyme B, and interleukin 1β converting enzyme.

The AMC labeled PS-SCLs clearly revealed the potential of PS-SCLs toward understanding protease specificity. However, they were also inherently limited in scope. Synthesis of an AMC peptide substrate requires the AMC leaving group to be attached to the P1 amino acid C-terminus prior to solid phase synthesis. Under normal solid phase peptide synthesis conditions, the C-terminal carboxyl group of this first amino acid would be used for coupling to the resin. However, the presence of AMC on the carboxyl terminus necessitates coupling to the resin through the amino acid side chain. This requirement severely limits the number of amino acids that can be incorporated into the P1 position to those with side chains that can be functionalized. Overall, the limitations produced AMC libraries that contained fixed P1 residues throughout while the P2–P4 residues were varied to test specificity. By restricting the P1 position to a single amino acid, these initial libraries were ideal for proteases with known P1 specificities, but had limited usefulness toward novel proteases.

The promise of these libraries coupled with this limitation led to the development of libraries utilizing a 7-amino-4-carbamoylmethylcoumarin (ACC) fluorogenic leaving group *(4, 5)*. The bifunctionality of ACC allows for the attachment of an ACC-Fmoc directly to solid phase resin using standard Fmoc chemistry. Furthermore, the ACC leaving group displays an approximately threefold higher fluorescent yield than AMC, increasing the sensitivity of the assay *(4)*. Once attached to the resin, Fmoc chemistry can be used to incorporate any amino acid at the initial (P1) position, allowing all of the peptide positions to be varied. This development led to the complete diverse ACC-tetrapeptide libraries, which are able to analyze each position of the peptide substrate independently. Most importantly, with the ability to completely vary the P1 position, proteases of unknown specificity could now be assayed.

The complete diverse ACC-tetrapeptide PS-SCL contains 160,000 different sequences that can be assayed in a simple, rapid, and easy to interpret format. In this library, each position of the tetrapeptide is held constant for every amino acid while the remaining three positions are randomized (**Fig. 1**). For each position, this creates 20 sublibraries, with each sublibrary composed of 8,000 different peptides. By only holding one position constant, the role of the constant residue can be independently analyzed. Because the remaining positions are randomized, the importance of any particular residue in these positions is masked. The overall result reveals the relative preference of the protease for amino acids at positions P1–P4, as shown in **Fig. 2** for the protease MT-SP1.

Although it is extremely useful in determining the preferred residues at P1 through P4 for a given protease, there is an important caveat to interpretation of the data. By randomizing every position other than the position of interest, each sublibrary contains all possible peptides and ensures that if the residue in the constant position supports cleavage, a subset will be cleaved. However, at the same time, this design masks any interdependence between positions. The appearance of incompatible amino acids in positions other than the one being held constant will not be apparent. Returning to the results for MT-SP1 (**Fig. 2**), the preferred substrate appears to be P4 Arg, P3 Lys, P2 Ser, and P1 Lys (RKSK). However, it is possible that each residue, although optimal when the other three positions are completely randomized, may not be compatible with one or more of the other residues when placed in the same sequence. For MT-SP1, this was exactly the case. An interdependence between binding pockets was revealed by phage display libraries that provided discrete peptide substrates to the protease *(6)*. Similar to the library

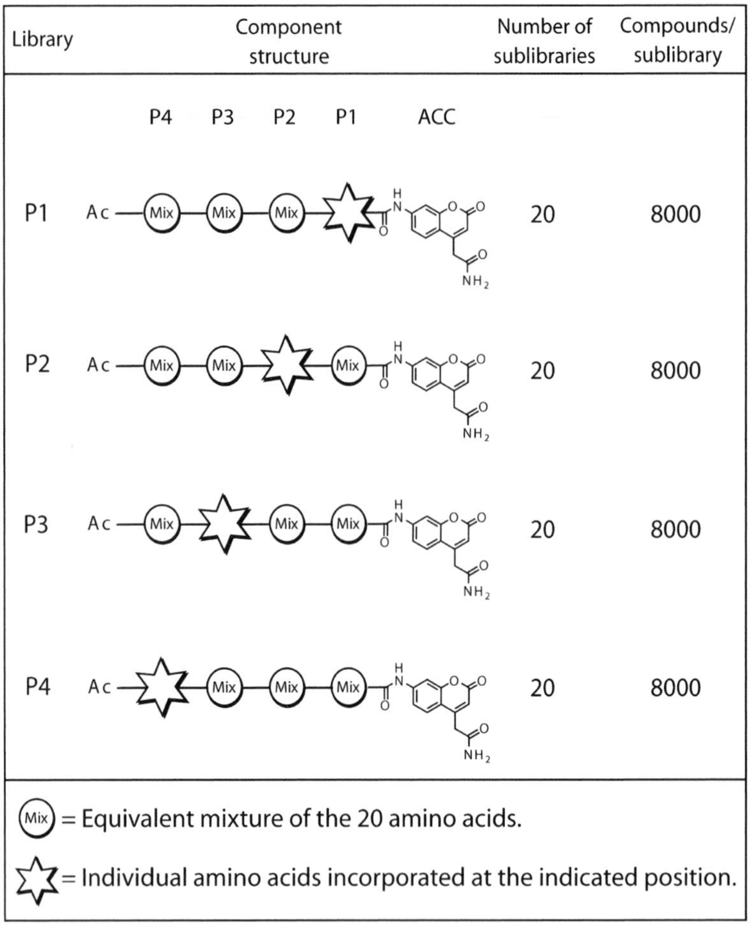

Fig. 1. Composition of the complete diverse ACC tetrapeptide PS-SCL libraries. Each library holds one position constant for each of the 20 amino acids while the remaining positions are randomized. This creates 20 sublibraries per library (one per amino acid), with each sublibrary a mixture of 8,000 peptides as a result of randomization at the other three positions (20^3). (Figure adapted from **ref**.8).

results, the P3 and P4 pockets were shown to prefer basic residues. However, both positions could not be a basic residue in a single peptide. Using the library data alone, this interdependence could not be detected at the P3 and P4 positions.

Alternatively, some proteases do not exhibit cooperativity between binding sites. In a study on granzymes A and B, the library results properly identified the optimal substrates for both, indicating that there was no interdependence between the binding sites (7). As seen in **Fig. 3**, the optimal substrate for granzyme B was identified as P4 Ile, P3 Glu, P2 Pro, and P1 Asp (IEPD). Furthermore, this granzyme B substrate was found to be more sensitive than any previously reported substrate. Because it is difficult to predict if the binding sites will exhibit interdependence,

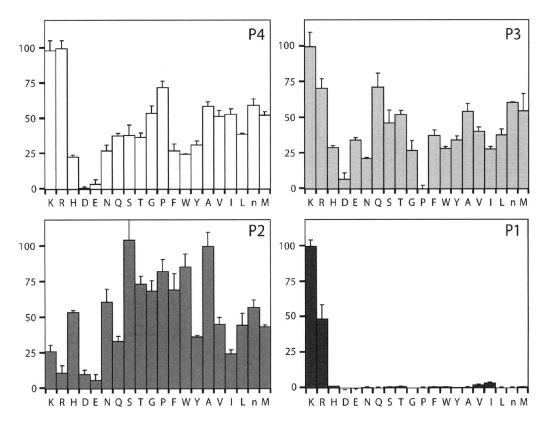

Fig. 2. Complete diverse ACC-tetrapeptide PS-SCL results for MT-SP1 for each library P1 through P4. The *y*-axis is the relative percent activity within each library. The *x*-axis indicates the amino acid being held constant.

like MT-SP1, or not, like granzymes A and B, the substrate candidate tetrapeptide sequences revealed through library profiling must be individually synthesized and tested to unequivocally show which one is optimal.

Use of the complete diverse ACC tetrapeptide PS-SCL has lead to an increased understanding of the P1 through P4 specificity for a number of proteases implicated in cancer. Furthermore, use of this information has led to the creation of specific inhibitors as well as fulfilling the ultimate goal of assisting in the identification of new in vivo substrates. Examples of these uses can be found in a number of recent papers. In a paper by Choe et al., results from the library were used to design unique tetrapeptide substrates specific for cathepsin K over cathepsins S, L V, and B *(8)*. In addition, one of these peptide sequences was incorporated into a mechanism based inhibitor of cathepsin K. Although not as selective as the original peptide substrate, there was no inhibition of cathepsins L and V, while inhibiting B with a k_{inact} about twofold slower than for cathepsin K. This indicates that results from the library may be utilized to design selective protease inhibitors.

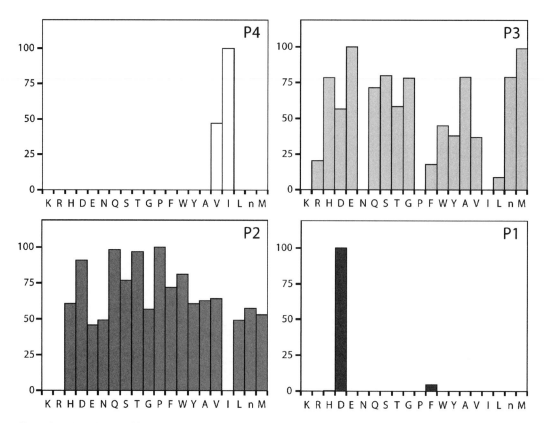

Fig. 3. Complete diverse ACC-tetrapeptide PS-SCL results for granzyme B for each library P1 through P4. The *y*-axis is the relative percent activity within each library. The *x*-axis indicates the amino acid being held constant.

In a separate paper, Marnett et al. analyzed the substrate specificity of Kaposi's sarcoma-associated herpesvirus (KSHV) protease with the complete diverse library *(9)*. The results revealed an unexpected preference for aromatic residues at P4. The complete results were successfully used to design both a tetrapeptide ACC substrate that was far superior to anything available as well as a mechanism based diphenyl phosphonate active site inhibitor. These tools helped to uncover a unique dimerization-dependent method of protease activation.

In a third example, the library was used by Bhatt et al. to assist in the identification of novel MT-SP1 substrates *(10)*. The data for MT-SP1 (similar to that shown above in **Fig. 1**) was combined with known endogenous substrate sequences and used to identify novel candidate substrates. Subsequent transcriptional profiling and cell-based assays confirmed the identification of MSP-1 and its receptor RON as a MT-SP1 substrate. This in turn revealed an extracellular proteolytic signaling pathway that is important for metastatic breast cancer *(11)*.

These results all demonstrate the usefulness of the complete diverse PS-SCL library in not only determining the substrate specificity of proteases, but also in the development of new tools for protease research. However, many proteases have additional specificity in the prime side pockets, which this library is unable to examine. Of particular importance to cancer research are the metalloproteases, which exhibit a large dependence upon prime side binding for activity. For this reason, the development of tools to target the prime side specificity of these proteases is ongoing. One method developed by Barrios et al. has shown promise, but suffers from a relatively high background *(12)*. More work is currently being carried out to develop a prime side library that demonstrates the robustness and utility of the complete diverse ACC-tetrapeptide PS-SCL described here.

The methods presented here begin with the synthesis of the P1 through P4 complete diverse ACC-tetrapeptide positional scanning synthetic combinatorial libraries. Each library is synthesized using standard Fmoc chemistry on an ACC linked resin to incorporate the fluorescent leaving group. Each library holds one of the positions of the tetrapeptide constant for all 20 amino acids, while completely randomizing the remaining 3 positions, as shown in the synthetic flow chart in **Fig. 4**. The result is 20 sublibraries (one for each amino acid) per position in the tetrapeptide. The protocol for testing proteases against the newly synthesized libraries is then presented. Finally, the synthesis of individual substrates for confirmation of the library data is described.

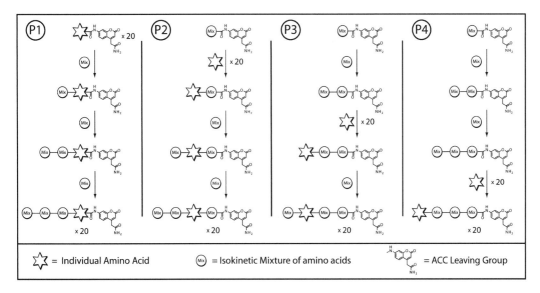

Fig. 4. Overview of the synthesis for each library of the Complete Diverse Tetrapeptide-ACC PS-SCL. The steps shown incorporate a specific amino acid in the position labeled with a star and an equal mixture of the 20 amino acids in the other three positions. The final result is one of the 20 sublibraries for each library. The steps are then repeated 19 more times to complete each library for all 20 amino acids.

2. Materials

2.1. Synthesis of PS-SCL ACC-Tetrapeptide Library

1. 7-Fmoc-aminocoumarin-4-acetic acid (*see* **Note 1**).
2. Double distilled water (H_2O).
3. Rink amide AM resin.
4. *N,N*-Dimethylformamide (DMF).
5. Piperidine.
6. *N*-Hydroxybenzotriazole (HOBt).
7. Diisopropylcarbodiimide (DICI).
8. Tetrahydrofuran (THF).
9. Phosphorous pentoxide (P_2O_5).
10. Methanol.
11. Fmoc amino acids (Fmoc-Ala-OH, Fmoc-Arg(Pbf)-OH, Fmoc-Asn(Trt)-OH, Fmoc-Asp(*O-t*-Bu)-OH, Fmoc-Glu (*O-t*-Bu)-OH, Fmoc-Gln(Trt)-OH, Fmoc-Gly-OH, Fmoc-His(Trt)-OH, Fmoc-Ile-OH, Fmoc-Leu-OH, Fmoc-Lys (Boc)-OH, Fmoc-Met-OH, Fmoc-Nle-OH, Fmoc-Phe-OH, Fmoc-Pro-OH, Fmoc-Ser(*O-t*-Bu)-OH, Fmoc-Thr(*O-t*-Bu)-OH, Fmoc-Trp(Boc)-OH, Fmoc-Tyr(*O-t*-Bu)-OH, Fmoc-Val-OH) (*see* **Note 2**).
12. Collidine.
13. 2-(1H-7-Azabenzotriazol-1-yl)-1,1,3,3-tetramethyl uronium hexafluorophosphate methanaminium (HATU).
14. Acetic acid.
15. Nitrotriazole.
16. Dichloromethane.
17. Trifluoroacetic acid.
18. Empty, fritted solid phase extraction (SPE) column.
19. Vacuum line with trap.
20. Rotary evaporator.
21. Argonaut Quest 210 organic synthesizer (Argonaut Technologies).
22. FlexChem 48-well reactor block (SciGene, Robbins Scientific).

2.2. Determination of Protease Specificity Using PS-SCL Library

1. P1, P2, P3, and P4 PS-SCL ACC libraries as synthesized in **Subheading 3.1.**
2. The protease of interest, typically less than 25 nmol but possibly more.
3. Optimal assay buffer (25 mL) for protease activity.

4. Fluorescence microplate reader (e.g. Molecular Devices SpectraMax Gemini)

5. Round-bottom 96-well plates for fluorescent plate readers (e.g., Thermo Scientific Microfluor 1 Black).

6. Multi-channel pipette capable of delivering 12 × 100 μL.

2.3. Synthesis of Individual Tetrapeptide ACC Substrates for Kinetic Verification

1. Empty, fritted SPE column.

2. DMF.

3. P1 Fmoc amino acids bound to ACC-resin (from 3.1.B).

4. Individual Fmoc amino acids.

5. Piperidine.

6. HOBt.

7. DICI.

8. Acetic acid.

9. Dichloromethane.

10. Trifluoroacetic acid (TFA).

11. Triisopropylsilane.

12. H_2O.

13. Dimethyl sulfoxide (DMSO).

3. Methods

3.1. Synthesis of the Complete Diverse PS-SCL ACC-Tetrapeptide Library

3.1.1. Acc-Rink Amide Resin

1. In an empty round bottom flask, swell 17 mmol (21 g) of Rink Amide AM resin with 200 mL N,N-dimethylformamide (DMF) for 30 min to make the functional groups accessible with gentle overhead stirring.

2. Vacuum filter to remove the DMF.

3. Add 200 mL of 20% piperidine in DMF to the resin to remove the Fmoc protecting group and activate the amine. Gently agitate to mix (*see* **Note 3**).

4. After 25 min, remove the piperidine by vacuum filtration and wash three times with 20 mL of DMF.

5. Add 2 equivalents (34 mmol, 15 g) of 7-Fmoc-aminocoumarin-4-acetic acid along with 2 equivalents N-hydroxybenzotriazole (HOBt) (34 mmol, 4.6 g) and 150 mL DMF followed by 2 equivalents diisopropylcarbodiimide (DICI) (34 mmol, 5.3 mL).

6. Mix with gentle agitation overnight.

7. Filter the resin mixture and wash three times with 200 mL DMF followed by three washes with 200 mL tetrahydrofuran and three washes with 200 mL methanol.

8. Dry over P_2O_5 under vacuum.

9. Determine the substitution level of the resin by Fmoc analysis following the Millipore procedure (*see* **Note 4**) *(13)*.

3.1.2. Preparation of Fmoc Amino Acid Substituted ACC-Resin

1. Add 100 mg of the Fmoc-ACC resin to 20 reaction vessels of an Argonaut Quest 210 organic synthesizer and swell with 2 mL of DMF (*see* **Note 5**). Calculate the molar quantity according to the substitution level determined above in **Subheading 3.1.1**, **step 9**.

2. Filter resin and add 2 mL of 20% piperidine in DMF to each reaction vessel. Gently agitate for 25 min to remove the Fmoc protecting group.

3. Filter resin and wash three times with 2 mL DMF.

4. To one of the reaction vessels containing the ACC resin add 5 molar equivalents of the Fmoc-Ala-OH along with 0.7 mL DMF, 10 molar equivalents of collidine, and 5 molar equivalents of HATU. Similarly add the remaining 19 Fmoc amino acids to the 19 wells containing ACC-resin (*see* **Note 6**). Gently agitate for 20 h.

5. Filter the resin and wash three times with 2 mL DMF.

6. Repeat **step 4** to ensure coupling occurs at a high efficiency.

7. Add 0.7 mL DMF, 40 µL acetic acid (0.7 mmol), 110 µL DICI (0.7 mmol), and 80 mg 3-nitro-1,2,4-triazol (0.7 mmol) to each reaction vessel and mix for 24 h to cap any unreacted ACC-resin.

8. Filter the resin and wash three times with 2 mL DMF.

9. Repeat wash three times with 2 mL tetrahydrofuran and three times with methanol.

10. Dry the Fmoc-amino acids bound to ACC-resin over P_2O_5 under vacuum.

11. Determine the substitution level by Fmoc analysis as in **Subheading 3.1.1**, **step 9**.

3.1.3. Synthesis of the P1 Sublibraries

1. According to the calculated Fmoc substitution levels, add 0.1 mmol of each of the 20 single Fmoc amino acids bound to ACC resin (as prepared in **Subheading 3.1.2**) to 20 separate wells of a MultiChem 48-well synthesis apparatus. Each of the 20 wells should contain a different Fmoc-amino acid bound to ACC-resin.

2. Add 4 mL of DMF to each well for 30 min to solvate the resin.

3. Remove the DMF and add 4 mL of 20% piperidine in DMF to each well and gently agitate for 30 min to remove the Fmoc protecting group.

4. Remove the piperidine solution and wash three times with 4 mL DMF.

5. Activate 20 mmol of an isokinetic mixture (*see* **Note 7**) of the 20 Fmoc amino acids (enough for all 20 wells at 10 equivalents/well compared to resin) in 80 mL DMF contianing 20 mmol HOBt and 20 mmol DICI.

6. Add 4 mL of the activated isokinetic Fmoc amino acid mixture to each of the 20 reaction vessels containing an Fmoc amino acid bound to ACC resin. Allow to couple with gentle agitation for 3 h. This will add a randomized P2 amino acid to each P1 residue.

7. Drain and wash the resin three times with 4 mL DMF.

8. Repeat steps 3–7 two more times to add a randomized P3 and P4 residue, creating the 20 P1 sublibraries.

9. Remove the final Fmoc protecting group with 4 mL of 20% piperidine in DMF with gentle agitation for 30 min.

10. Remove the piperidine solution and wash three times with 4 mL DMF.

11. Cap the final peptide with 80 mmol acetic acid, 80 mmol HOBt, and 80 mmol DICI in 4 mL DMF for 4 h with gentle agitation

12. Remove the capping solution and wash three times with 4 mL DMF followed by three washes with 4 mL dichloromethane.

13. Cleave the peptides from the resin in a solution of 2,850 μL trifluoroacetic acid (TFA), 75 μL triisopropylsilane, and 75 μL water (95:2.5:2.5) for 1 h with gentle agitation.

14. Collect the cleaved peptides and lyophilize.

15. Dissolve the peptides in dimethyl sulfoxide (DMSO) to a final concentration of 25 mM (*see* **Note 8**).

3.1.4. Synthesis of the P2 Sublibraries

1. According to the calculated substitution levels, create a 2 mmol mixture containing equal amounts of the 20 Fmoc-amino acids bound to ACC-resin (0.1 mmol each) and combine with 4 mL DMF in an empty SPE column. Mix for 2 h with gentle agitation.

2. Remove the DMF and dry the resin mixture.

3. Split the resin evenly between 20 reaction vessels of a MultiChem 48-well synthesis apparatus (0.1 mmol resin per reaction vessel).

4. Add 4 mL DMF for 30 min to solvate the resin mixtures.

5. Remove the DMF and add 4 mL of 20% piperidine in DMF to each well and gently agitate for 30 min to remove the Fmoc protecting group.

6. Remove the piperidine solution and wash three times with 4 mL DMF.

7. To insert individual amino acids in the P2 position, separately activate 1 mmol (10 molar equivalents compared to the resin) of each of the 20 Fmoc-amino acids in 4 mL DMF containing 1 mmol HOBt and 1 mmol DICI.

8. Add one of the activated Fmoc amino acids to one of the 20 reaction vessels containing the mixture of P1 amino acids bound to ACC-resin. Similarly add the 19 remaining activated Fmoc amino acids to the 19 reaction vessels containing resin.

9. Allow to couple with gentle agitation for 3 h.

10. Drain and wash the resin three times with 4 mL DMF.

11. Add 4 mL of 20% piperidine in DMF to each well and gently agitate for 30 min to remove the Fmoc protecting group.

12. Remove the piperidine solution and wash three times with 4 mL DMF

13. Activate 20 mmol of an isokinetic mixture of the 20 Fmoc amino acids (enough for all 20 wells at 10 equivalents/well compared with resin) with 20 mmol HOBt and 20 mmol DICI in 80 mL DMF.

14. Add 4 mL of the activated isokinetic Fmoc amino acid mixture to each reaction vessel containing resin. Couple with agitation for 3 h. This will add a randomized P3 residue to each known P2.

15. Drain and wash the resin three times with 4 mL DMF.

16. Repeat **steps 11–15** once more to add a randomized P4 residue to each peptide, creating the P2 sublibraries.

17. Remove the final Fmoc protecting group with 4 mL of 20% piperidine in DMF with gentle agitation for 30 min.

18. Remove the piperidine solution and wash three times with 4 mL DMF.

19. Cap the final peptide with 80 mmol acetic acid, 80 mmol HOBt, and 80 mmol DICI in 4 mL DMF for 4 h.

20. Remove the capping solution and wash three times with 4 mL DMF followed by three washes with 4 mL dichloromethane.

21. Cleave the peptides from the resin in a solution of 2,850 μL TFA, 75 μL triisopropylsilane, and 75 μL water (95:2.5:2.5) for 1 h with gentle agitation.

22. Collect the cleaved peptides and lyophilize.

23. Dissolve the peptides in DMSO to a final concentration of 25 mM.

3.1.5. Synthesis of the P3 Sublibraries

1. As in **Subheading 3.1.4, step 1**, create a 2 mmol mixture of the 20 Fmoc-amino acids bound to ACC-resin and add it to 4 mL DMF in an empty SPE column. Mix for 2 h with gentle agitation.

2. Remove the DMF and dry the resin mixture.

3. Split the resin evenly between 20 reaction vessels of a MultiChem 48-well synthesis apparatus (0.1 mmol resin per each vessel).

4. Add 4 mL DMF for 30 min to solvate the resin mixtures.

5. Remove the DMF and add 4 mL of 20% piperidine in DMF to each well and gently agitate for 30 min to remove the Fmoc protecting group.

6. Remove the piperidine solution and wash three times with 4 mL DMF.

7. Activate 20 mmol of an isokinetic mixture of the 20 Fmoc amino acids (enough for all 20 wells at 10 equivalents/well compared to resin) with 20 mmol HOBt and 20 mmol DICI in 80 mL DMF.

8. Add 4 mL of the activated isokinetic Fmoc amino acid mixture to each reaction vessel containing the mixture of Fmoc amino acids bound to ACC resin. Allow to couple with gentle agitation for 3 h. This will add a randomized P2 residue.

9. Drain and wash the resin three times with 4 mL DMF.

10. Add 4 mL of 20% piperidine in DMF to each well and gently agitate for 30 min to remove the Fmoc protecting group.

11. Remove the piperidine solution and wash three times with 4 mL DMF.

12. To insert individual amino acids in the P3 position, separately activate 1 mmol (10 molar equivalents compared to the resin) of each of the 20 Fmoc-amino acids in 4 mL DMF containing 1 mmol HOBt and 1 mmol DICI.

13. Add one of the activated Fmoc amino acids to one of the 20 reaction vessels containing resin. Similarly add the other 19 activated Fmoc amino acids to the 19 remaining reaction vessels containing resin.

14. Allow to couple with gentle agitation for 3 h.

15. Drain and wash the resin three times with 4 mL DMF.

16. Repeat steps 4–9 once to add a randomized P4 residue to each peptide, creating the P3 sublibraries.

17. Remove the final Fmoc protecting group with 4 mL of 20% piperidine in DMF with agitation for 30 min.

18. Remove the piperidine solution and wash three times with 4 mL DMF.

19. Cap the final peptide with 80 mmol acetic acid, 80 mmol HOBt, and 80 mmol DICI in 4 mL DMF for 4 h with gentle agitation.

20. Remove the capping solution and wash three times with 4 mL DMF followed by three washes with 4 mL dichloromethane.

21. Cleave the peptides from the resin in a solution of 2,850 µL TFA, 75 µL triisopropylsilane, and 75 µL water (95:2.5:2.5) for 1 h with gentle agitation.

22. Collect the cleaved peptides and lyophilize.

23. Dissolve the peptides in DMSO to a final concentration of 25 mM.

3.1.6. Synthesis of the P4 Sublibraries

1. As in **Subheading 3.1.4**, **step 1**, create a 2 mmol mixture of the 20 Fmoc-amino acids bound to ACC-resin and add it to 4 mL DMF in an empty SPE column. Mix for 2 h with gentle agitation.

2. Remove the DMF and dry the resin mixture.

3. Split the resin evenly between 20 reaction vessels of a MultiChem 48-well synthesis apparatus (0.1 mmol resin per each vessel).

4. Add 4 mL DMF for 30 min to solvate the resin mixtures.

5. Remove the DMF and add 4 mL of 20% piperidine in DMF to each well and gently agitate for 30 min to remove the Fmoc protecting group.

6. Remove the piperidine solution and wash three times with 4 mL DMF.

7. Activate 20 mmol of an isokinetic mixture of the 20 Fmoc amino acids (enough for all 20 wells at 10 equivalents/well compared to resin) with 20 mmol HOBt and 20 mmol DICI in 80 mL DMF.

8. Add 4 mL of the activated isokinetic Fmoc amino acid mixture to each reaction vessel containing the mixture of Fmoc amino acids bound to ACC resin. Allow to couple with gentle agitation for 3 h. This will add a randomized P2 residue.

9. Drain and wash the resin three times with 4 mL DMF.

10. Repeat **steps 5–9** to add a randomized P3 residue.

11. Add 4 mL of 20% piperidine in DMF to each well and gently agitate for 30 min to remove the Fmoc protecting group.

12. Remove the piperidine solution and wash three times with 4 mL DMF.

13. To insert individual amino acids in the P4 position, separately activate 1 mmol (10 molar equivalents compared to the resin) of each of the 20 Fmoc-amino acids in 4 mL DMF containing 1 mmol HOBt and 1 mmol DICI.

14. Add one of the activated Fmoc amino acids to one of the 20 reaction vessels containing resin. Similarly add the 19 remaining activated Fmoc amino acids to the 19 reaction vessels containing resin.

15. Couple with gentle agitation for 3 h.

16. Drain and wash the resin three times with 4 mL DMF.

17. Remove the final Fmoc protecting group with 4 mL of 20% piperidine in DMF with agitation for 30 min.

18. Remove the piperidine solution and wash three times with 4 mL DMF.

19. Cap the final peptide with 80 mmol acetic acid, 80 mmol HOBt, and 80 mmol DICI in 4 mL DMF for 4 h.

20. Remove the capping solution and wash three times with 4 mL DMF followed by three washes with 4 mL dichloromethane.

21. Cleave the peptides from the resin in a solution of 2,850 μL TFA, 75 μL triisopropylsilane, and 75 μL water (95:2.5:2.5) for 1 h with agitation.

22. Collect the cleaved peptides and lyophilize.

23. Dissolve the peptides in DMSO at a final concentration of 25 mM.

3.2. Determination of Protease Specificity Using the Complete Diverse ACC-Tetrapeptide PS-Scls

1. To ensure that the protease will exhibit high enough activity to observe cleavage throughout the library, the activity of the enzyme should be tested at various concentrations against at least one of the P1 sublibraries that correspond to a known, or expected, P1 preferred residue for the protease. This information can then be used to adjust the protease concentration to give sufficient activity in the assay.

2. To test the activity, prepare protease samples at various concentrations within the typical working range of 1 to 1,000 nM in 100 μL assay buffer optimized for the protease activity.

3. Spot 1 μL of the selected 25 mM P1 sublibrary in wells of a 96-well plate (*see* **Note 9**) (e.g., MT-SP1 (shown in **Fig. 2**) has known trypsin-like activity and would therefore be initially tested against the P1 Arg or Lys sublibrary.)

4. Add the 100 μL protease samples prepared at various concentrations to the wells containing the sublibrary sample.

5. Monitor the hydrolysis reaction on a fluorescence plate reader taking readings every 15 s for 30 min with excitation at 380 nm and emission at 460 nm (*see* **Note 10**).

6. A good working concentration will give an rfu/sec reading of between 1 and 10 for a known substrate P1 residue. This will provide enough activity to measure cleavage of substrates containing less optimal residues at the constant position. Adjust the protease concentration appropriately to have activity within this range.

7. To keep the results consistent, it is best to prepare enough protease to run the entire library, P1 through P4. When performed in triplicate, the assay requires 24 mL of protease in assay buffer (triplicate × 20 residues/library × 100 μL/assay × 4 libraries = 24,000 μL).

8. To ensure enough sample for the entire assay, prepare > 24 mL of the protease in assay buffer at the enzyme concentration determined in steps 1–6.

9. In a 96-well plate, spot 1 μL of each P1 sublibrary substrate in triplicate, giving a total of 60 reactions.

10. Using a multi-channel pipette, add 100 μL of the protease solution to each of the 60 wells.

11. Monitor the hydrolysis reaction on the fluorescence plate reader taking readings every 15 s for 30 min with excitation at 380 nm and emission at 460 nm.

12. Repeat steps 8–11 for the P2, P3, and P4 libraries.

13. The results (reaction velocities) for each library can be standardized to the residue within that library giving the maximum activity for general comparison, as shown in Fig. 2 (*see* **Note 11**).

3.3. Synthesis of Individual Tetrapeptide ACC Substrates for Kinetic Verification

1. Because the results from the complete diverse ACC PS-SCL do not take into consideration the effects that neighboring residues contribute to binding pocket specificity, it is important to verify the preferred sequence suggested by the library. To test this, a few of the preferred sequences should be synthesized as ACC substrates following a procedure similar to that outlined for each sublibrary.

2. According to the calculated Fmoc substitution levels, add 0.1 mmol of the desired P1 Fmoc amino acid bound to ACC resin to an empty fritted SPE column.

3. Add 4 mL of DMF and gently agitate for 30 min to solvate the resin with.

4. Remove the DMF and add 4 mL of 20% piperidine in DMF and gently agitate for 30 min to remove the Fmoc protecting group.

5. Remove the piperidine solution and wash three times with 4 mL DMF.

6. Activate 1 mmol of the desired P2 Fmoc amino acid (10 equivalents compared with the P1 substituted ACC resin) in 4 mL DMF with 1 mmol HOBt and 1 mmol DICI.

7. Add the activated Fmoc amino acid to the reaction vessel containing the P1 amino acid bound to ACC resin. Allow to couple with gentle agitation for 3 h.

8. Drain and wash the resin three times with 4 mL DMF.

9. Repeat steps 4–8 two more times to add the desired P3 and P4 residues to the peptide.

10. Remove the final Fmoc protecting group with 4 mL of 20% piperidine in DMF with gentle agitation for 30 min.

11. Remove the piperidine solution and wash three times with 4 mL DMF.

12. Cap the final peptide with 80 mmol acetic acid, 80 mmol HOBt, and 80 mmol DICI in 4 mL DMF for 4 h.

13. Remove the capping solution and wash three times with 4 mL DMF followed by three washes with 4 mL dichloromethane.

14. Cleave the peptide from the resin in a solution of 2,850 μL TFA, 75 μL triisopropylsilane, and 75 μL H_2O (95:2.5:2.5) for 1 h with gentle agitation.

15. Collect the cleaved peptide and lyophilize.

16. Dissolve the peptide in DMSO to the desired concentration for kinetic evaluation.

4. Notes

1. 7-Fmoc-aminocoumarin-4-acetic acid can be synthesized according to the procedures outlined by Harris et al. *(4)*, Kanaoka et al. *(14)*, Besson et al. *(15)*.

2. Cysteine was excluded due to its propensity to polymerize. Norleucine was included to extend the information revealed by the library. Norleucine has a four carbon side-chain similar to leucine and isoleucine, but unlike those amino acids it is unbranched. This extended form offers additional information about the protease binding pocket.

3. It is important not to use a stirbar for agitation of the resin to avoid damaging the resin. A platform shaker modified to hold the fritted reaction or the FlexChem reactor block

upright works well for the gentle agitation required throughout this method. Alternatively an overhead stirrer could be used in some instances.

4. The method for Fmoc analysis is as follows, taken from Bunin *(13)*:Place a known quantity of resin in a 10 mL volumetric flask. Add 0.4 mL piperidine and 0.4 mL dichloromethane. After 30 min, add 1.6 mL of methanol and 7.6 mL dichloromethane to bring the volume to 10 mL. Measure the absorbance at 301 nm against a blank composed of 0.4 mL piperidine, 1.6 mL methanol, and 8.0 mL dichloromethane. Use the following equation to calculate the loading level:

$$Loading(mmol/g) = A_{301} \times 10mL/7,800 \times re\sin wt,$$

where A_{301} is the absorbance measured at 301 nm and 7,800 is the extinction coefficient of the piperidine-fluorenone adduct.

5. Addition of the first amino acid to the ACC-resin can also be accomplished in SPE columns following the same procedure. The Argonaut Quest was used to simplify the process.

6. The 20 amino acids used in the library described here are the 20 common amino acids excluding cysteine and including norleucine.

7. An isokinetic mixture contains a mixture of amino acids at a ratio based on the individual reaction kinetics of coupling with a free amine that ensures an equal distribution of each amino acid in the final product. This is based on the findings that individual amino acids react with the free amine of a resin bound amino acid at different rates, but independent of the identity of the resin bound amino acid *(4, 16)*. To create the library described (excluding cysteine and including norleucine) the following ratio was used (Amino Acid, mol %): Fmoc-Ala-OH, 3.3; Fmoc-Arg(Pbf)-OH, 6.4; Fmoc-Asn(Trt)-OH, 5.2; Fmoc-Asp(*O-t*-Bu)-OH, 3.4; Fmoc-Glu(*O-t*-Bu)-OH, 3.5; Fmoc-Gln(Trt)-OH, 5.2; Fmoc-Gly-OH, 2.8; Fmoc-His(Trt)-OH, 3.4; Fmoc-Ile-OH, 17.0; Fmoc-Leu-OH, 4.8; Fmoc-Lys(Boc)-OH, 6.1; Fmoc-Met-OH, 2.2; Fmoc-Nle-OH, 3.7; Fmoc-Phe-OH, 2.4; Fmoc-Pro-OH, 4.2; Fmoc-Ser(*O-t*-Bu)-OH, 2.7; Fmoc-Thr(*O-t*-Bu)-OH, 4.7; Fmoc-Trp(Boc)-OH, 3.7; Fmoc-Tyr(*O-t*-Bu)-OH, 4.0; Fmoc-Val-OH, 11.1. These isokinetic ratios were prepared based on those reported by Harris et al. and Ostresh et al. *(4, 16)*.

8. After solubilizing in DMSO, the libraries should be stored at −80°C.

9. Some of the peptide sublibraries are less soluble than others and may require gentle warming to ensure they are completely dissolved. Prior to use, check that all peptides are completely in solution and warm slightly if necessary.

10. With some proteases the resulting fluorescence readout jumps up and down over time, with an overall upward slope. Many times this can be reduced by addition of 0.01% Brij 35 to the reaction buffer.

11. Alternatively, the rfu/sec data can be converted to pM/sec and reported. This conversion factor must be experimentally determined for each instrument by measuring the rfu resulting from the complete hydrolysis of a known quantity of ACC-peptide. For the Molecular Devices SpectraMax Gemini, we have measured 1 pM of hydrolyzed ACC is equal to 1,380 rfu.

Acknowledgments

We thank M. D. Lim, M. Zhao, and Y. Choe for helpful discussions and critical reading of the manuscript. The authors were funded by NIH grants CA108462-04 (E.L.S.) and GM082250 (C.S.C).

References

1. Houghten, R. A., Pinilla, C., Blondelle, S. E., Appel, J. R., Dooley, C. T., and Cuervo, J. H. (1991) Generation and use of synthetic peptide combinatorial libraries for basic research and drug discovery. *Nature* 354, 84–6.

2. Rano, T. A., Timkey, T., Peterson, E. P., Rotonda, J., Nicholson, D. W., Becker, J. W., Chapman, K. T., and Thornberry, N. A. (1997) A combinatorial approach for determining protease specificities: application to interleukin-1beta converting enzyme (ICE). *Chem Biol* 4, 149–55.

3. Thornberry, N. A., Rano, T. A., Peterson, E. P., Rasper, D. M., Timkey, T., Garcia-Calvo, M., Houtzager, V. M., Nordstrom, P. A., Roy, S., Vaillancourt, J. P., Chapman, K. T., and Nicholson, D. W. (1997) A combinatorial approach defines specificities of members of the caspase family and granzyme B. Functional relationships established for key mediators of apoptosis. *J Biol Chem* 272, 17907–11.

4. Harris, J. L., Backes, B. J., Leonetti, F., Mahrus, S., Ellman, J. A., and Craik, C. S. (2000) Rapid and general profiling of protease specificity by using combinatorial fluorogenic substrate libraries. *Proc Natl Acad Sci USA* 97, 7754–9.

5. Maly, D. J., Leonetti, F., Backes, B. J., Dauber, D. S., Harris, J. L., Craik, C. S., and Ellman, J. A. (2002) Expedient solid-phase synthesis of fluorogenic protease substrates using the 7-amino-4-carbamoylmethylcoumarin (ACC) fluorophore. *J Org Chem* 67, 910–5.

6. Takeuchi, T., Harris, J. L., Huang, W., Yan, K. W., Coughlin, S. R., and Craik, C. S. (2000) Cellular localization of membrane-type serine protease 1 and identification of protease-activated receptor-2 and single-chain urokinase-type plasminogen activator as substrates. *J Biol Chem* 275, 26333–42.

7. Mahrus, S., and Craik, C. S. (2005) Selective chemical functional probes of granzymes A and B reveal granzyme B is a major effector of natural killer cell-mediated lysis of target cells. *Chem Biol* 12, 567–77.

8. Choe, Y., Leonetti, F., Greenbaum, D. C., Lecaille, F., Bogyo, M., Bromme, D., Ellman, J. A., and Craik, C. S. (2006) Substrate profiling of cysteine proteases using a combinatorial peptide library identifies functionally unique specificities. *J Biol Chem* 281, 12824–32.

9. Marnett, A. B., Nomura, A. M., Shimba, N., Ortiz de Montellano, P. R., and Craik, C. S. (2004) Communication between the active sites and dimer interface of a herpesvirus protease revealed by a transition-state inhibitor. *Proc Natl Acad Sci USA* 101, 6870–5.

10. Bhatt, A. S., Welm, A., Farady, C. J., Vasquez, M., Wilson, K., and Craik, C. S. (2007) Coordinate expression and functional profiling

identify an extracellular proteolytic signaling pathway. *Proc Natl Acad Sci USA* 104, 5771–6.

11. Welm, A. L., Sneddon, J. B., Taylor, C., Nuyten, D. S., van de Vijver, M. J., Hasegawa, B. H., and Bishop, J. M. (2007) The macrophage-stimulating protein pathway promotes metastasis in a mouse model for breast cancer and predicts poor prognosis in humans. *Proc Natl Acad Sci USA* 104, 7570–5.

12. Barrios, A. M., and Craik, C. S. (2002) Scanning the prime-site substrate specificity of proteolytic enzymes: a novel assay based on ligand-enhanced lanthanide ion fluorescence. *Bioorg Med Chem Lett* 12, 3619–23.

13. Bunin, B. A. (1998) *The Combinatorial Index*, Academic Press, San Diego.

14. Kanaoka, Y., Kobayashi, A., Sato, E., Nakayama, H., Ueno, T., Muno, D., and Sekine, T. (1984) Organic Fluorescent Reagents.10. Multifunctional Cross-Linking Reagents.1. Synthesis And Properties Of Novel Photoactivable, Thiol-Directed Fluorescent Reagents. *Chem Pharm Bull* 32, 3926–3933.

15. Besson, T., Joseph, B., Moreau, P., Viaud, M. C., Coudert, G., and Guillaumet, G. (1992) Synthesis and fluorescent properties of new heterobifunctional fluorescent-probes. *Heterocycles* 34, 273–291.

16. Ostresh, J. M., Winkle, J. H., Hamashin, V. T., and Houghten, R. A. (1994) Peptide libraries: determination of relative reaction rates of protected amino acids in competitive couplings. *Biopolymers* 34, 1681–9.

Chapter 5

Mixture-Based Peptide Libraries for Identifying Protease Cleavage Motifs

Benjamin E. Turk

Summary

All proteases and peptidases are to some extent sequence-specific, in that one or more residues are preferred at particular positions surrounding the cleavage site in substrates. I describe here a general protocol for determining protease cleavage site preferences using mixture-based peptide libraries. Initially a completely random, amino-terminally capped peptide mixture is digested with the protease of interest, and the cleavage products are analyzed by automated Edman sequencing. The distribution of amino acids found in each sequencing cycle indicates which residues are preferred by the protease at positions downstream of the cleavage site. On the basis of these results, a second peptide library is designed that is partially degenerate and partially fixed sequence. Edman sequencing analysis of the cleavage products of this peptide mixture provides preferences amino-terminal to the scissile bond. As necessary, the process is reiterated until the full cleavage motif of the protease is known. Cleavage specificity data obtained with this method have been used to generate specific and efficient peptide substrates, to design potent and specific inhibitors, and to identify novel protease substrates.

Key words: Peptide libraries, Proteases, Enzyme specificity, High-throughput screening, Edman sequencing.

1. Introduction

Proteolytic enzymes are encoded by some 2% of most genomes, reflecting their diverse roles in myriad biological processes *(1, 2)*. Although some proteases are digestive enzymes, many have highly specific protein processing functions. For both types of protease, substrate recognition is mediated in some part by interactions between the active site of the protease and the cleavage site

Toni M. Antalis and Thomas H. Bugge (eds.), *Methods in Molecular Biology, Proteases and Cancer, Vol. 539*
© Humana Press, a part of Springer Science+Business Media, LLC 2009
DOI: 10.1007/978-1-60327-003-8_5

within its substrate. Each protease has a unique cleavage motif, those amino acid residues that are preferred or required at specific positions surrounding the cleavage site. An understanding is that the sequence requirements for cleavage by a protease can be applied in a number of ways. For example, this information enables the design and production of specific and efficient peptide substrates that can be used for in vitro assay of the protease, and can be incorporated into protease biosensors that can be used in vivo or in cell culture *(3–5)*. In addition, peptide substrate analogs bearing chemical groups that target a particular class of protease can make highly potent inhibitors. Such inhibitors are useful both for biological studies and as leads in drug discovery. Lastly, proteins having sequences that match the cleavage motif can be identified by computer database searching, and these represent potential substrates for the protease *(6, 7)*.

Determining the cleavage motif for a protease requires analysis of a very large number of potential sequences: a ten amino acid stretch of polypeptide, for example, represents approximately 1×10^{13} possible combinations. A number of synthetic and encoded peptide library approaches have been developed to allow for parallel analysis of such large numbers of sequences *(8–12)*. The method I describe here (outlined in **Fig. 1**) involves Edman sequencing analysis of the cleavage products of peptide mixtures *(6, 13)*. The approach is iterative in that it involves using peptide libraries of gradually decreasing complexity. The

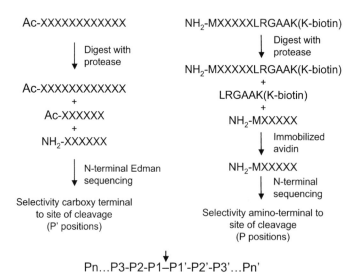

Fig. 1. Overview of the peptide library method. Initially a fully degenerate peptide mixture is used to elucidate preferences carboxy-terminal to the cleavage site (*left*). A partially degenerate peptide library designed on the basis of results from the initial screen is used to determine preferences amino-terminal to the cleavage site (*right*). As necessary, further rounds of iteration are performed. Figure adapted from ref. *6*.

initial peptide library is a completely random peptide mixture, typically a dodecamer that is capped at the amino-terminus with an acetyl group. The mixture is subjected to limited digestion with the protease of interest so that only those peptides within the pool that are the most efficient substrates are cleaved. The entire mixture is then subjected to automated amino-terminal Edman sequencing. Uncleaved peptides and the amino-terminal product fragments remain blocked and thus are not detected by the sequencer; only the carboxy-terminal fragments of the cleaved peptides are sequenced. The sequencer thus provides the distribution of amino acids found at positions carboxy-terminal to the cleavage site. The first sequencing cycle is composed of residues found at the P1′ position (immediately carboxy-terminal to the cleavage site, *see* **Note 1**), the second sequencing cycle corresponds to the P2′ position, and so on. Raw data are corrected for bias present within the starting mixture and normalized (*see* **Fig. 2**). A less complex secondary peptide library is then designed in which the most highly selected residues found at the primed positions are fixed, and a smaller number (typically five) positions are left degenerate. The secondary peptide library has a free amino terminus, and is biotinylated at its carboxy terminus. This peptide mixture is again partially digested with the protease, and the digest is treated in batch with immobilized avidin. Full length undigested peptides and carboxy-terminal fragments bind to the resin through the biotin tag, but amino-terminal cleavage products remain free. Edman sequencing of these fragments provides the sequence preference for the protease amino-terminal to the cleavage site. In some cases another round of iteration is required, and a tertiary peptide library is produced and analyzed in a manner similar to the secondary library.

This method allows one to quickly determine cleavage preferences for a given protease both upstream and downstream of the cleavage site. An advantage of this method over others is that it requires relatively few peptide mixtures. Both peptide synthesis and Edman sequencing can be done at reasonable cost by a facility, so little in the way of specialized equipment or expertise is required. A potential difficulty with the method is that determination of specificity at the unprimed positions requires that one be able to produce a secondary or tertiary library that is capable of directing cleavage by the protease largely (>90%) to a single peptide bond. This is generally trivial for proteases that are most highly selective at positions downstream of the cleavage site, as is the case for most metalloproteases. Serine and cysteine proteases, which tend to be most selective upstream of the cleavage site, can be more of a challenge. However, serine *(14)*, cysteine *(15)*, and metalloproteinases *(6, 16–18)* have all been successfully profiled by Edman sequencing of peptide mixture libraries.

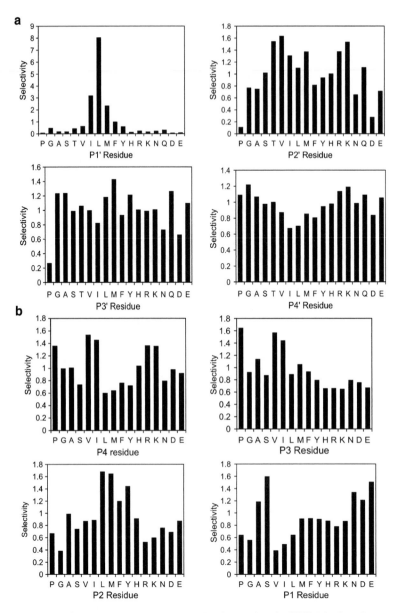

Fig. 2. Peptide library data: Corrected and normalized peptide library data for MMP-7 both carboxy-terminal (**a**) and amino-terminal (**b**) to the cleavage site are shown. The following peptide libraries were used to collect data: Ac-X_{12}-NH$_2$ (all primed positions), M-A-X-X-X-X-X-L-R-G-A-A-R-E-K(biotin) (P3 position), and M-G-X-X-P-X-X-L-R-G-G-G-E-E-K(biotin) (all other unprimed positions). Data were taken from ref. *(6)*.

2. Materials

2.1. Peptide Libraries

1. Random dodecamer peptide library, acetyl-X-X-X-X-X-X-X-X-X-X-X-X-amide (Ac-X_{12}-NH$_2$), where X is an equimolar mixture of the 19 naturally occurring amino acids excluding cysteine (custom made at a peptide synthesis facility, *see* **Note 2**),

2.7 mg/mL in 25 mM sodium phosphate, pH 7.4 (or 25 mM HEPES, pH 7.4 if protease is not active in phosphate buffer). Store frozen at –20°C in aliquots.

2. Uncapped random dodecamer peptide library, X-X-X-X-X-X-X-X-X-X-X-X-amide (X_{12}-NH_2), 2.7 mg/mL in the same buffer used for the acetylated library above. Store frozen at –20°C in aliquots.

3. Secondary peptide library with C-terminal biotinyl-lysine residue, 4 mg/mL in 25 mM phosphate, pH 7.4. Example: M-A-X-X-X-X-X-L-R-G-A-A-R-E-K(biotin)-amide. Store frozen at –20°C in aliquots.

4. Amino-terminally acetylated secondary peptide library, 4 mg/mL in 25 mM phosphate, pH 7.4. Store frozen at –20°C in aliquots.

5. Monomeric avidin (Pierce).

6. 2 mM biotin in phosphate buffered saline (PBS), prepared fresh.

7. 50 mM NH_4OAc (prepared fresh).

8. 0.2 M HOAc, store at room temperature for several months.

9. PBS containing 0.02% sodium azide (store at 4°C).

10. pH paper, range 5.0–10.0.

2.2. Protease Reactions

1. Purified protease of interest, catalytically active (*see* **Note 3**).

2. 10× reaction buffer, optimized for protease of interest (*see* **Note 4**).

3. 0.1 M HCl.

4. 0.1 M NaOH.

5. Fluorescamine (Sigma), 0.5 mg/mL in acetone. Can keep at room temperature for several weeks.

6. Avidin agarose (Sigma).

7. 50 mM NH_4HCO_3 (prepared fresh).

3. Methods

3.1. Determining Enzyme Concentration and Incubation Time

Before screening the peptide library, it is important to establish conditions under which approximately 5–10% of the peptides within the mixture are cleaved by the protease. Overdigestion will result in less specificity, since suboptimal sequences will be cleaved, while underdigestion will result in an insufficient signal over background. Optimal conditions are established by following cleavage of the peptide mixture over time using fluorescamine, a reagent that reacts with primary amines to give a stable fluorescent product *(19)*. Because the peptide mixtures contain lysine residues that react with fluorescamine, the background signal will prevent accurate quantitation if less than 10% of the peptides are cleaved. The optimal enzyme concentration and incubation times are determined by identifying conditions that result in a larger fraction cleaved, and then extrapolating to a lower enzyme concentration. After an initial Edman sequencing run (described in **Subheading 3.2** later), the enzyme concentration can be more finely adjusted based on the total cleaved peptide yield on the sequencer.

1. Prepare reaction tubes as follows:

    35 µL 2.7 mg/mL peptide library

    5 µL ddH$_2$O

    5 µL of 10× reaction buffer (optimized for protease of interest)

    5 µL protease, diluted in reaction buffer

    A total of six tubes should be prepared:

    A. Ac-X$_{12}$-NH$_2$ as the peptide library with no protease added

    B. Ac-X$_{12}$-NH$_2$ as the peptide library with 40 nM protease

    C. Ac-X$_{12}$-NH$_2$ as the peptide library with 200 nM protease

    D. Ac-X$_{12}$-NH$_2$ as the peptide library with 1 µM protease

    E. 1 µM protease without peptide to determine background signal from protease

    F. The uncapped peptide library X$_{12}$-NH$_2$ with no protease, as a standard

2. Incubate tubes at 37°C. At 30 min, 1, 2, and 4 h after starting the incubation, remove 10 µL aliquots of each reaction and transfer to fresh tubes containing 25 µL 0.1 M HCl to quench.

3. When the incubation is over, neutralize each quenched aliquot with 25 µL 0.1 M NaOH, and then dilute with 0.3 mL 25 mM HEPES, pH 7.4.

4. With vortexing, add 80 µL 0.5 mg/mL fluorescamine to each tube.

5. Read the fluorescence for each sample using an excitation wavelength of 387 nm and an emission wavelength of 480 nm.

6. Determine the extent of cleavage using the uncapped peptide library as a standard for 100% digestion, subtracting out the values from the protease alone and from the acetylated peptide library. Estimate the amount of enzyme required to provide 5% cleavage with a 2-h incubation time, assuming a linear relationship between enzyme concentration and reaction rate. If no cleavage can be detected above background under any conditions, an overnight digest with at 1 µM protease is recommended as the initial condition for peptide library screening.

3.2. Cleavage and Analysis of Primary Peptide Library

1. Set up a reaction tube containing:

   14 μL Ac-X$_{12}$-NH$_2$

   2 μL ddH2O

   2 μL of 10× reaction buffer (*see* **Note 4**)

   2 μL of 10× protease, diluted in reaction buffer

   A concentration of protease sufficient for cleavage of around 5% of the peptides within the mixture should be used.

2. Incubate the reaction for 2 h at 37°C, and then heat to 100°C for 2 min to stop the reaction.

3. Load 10 μL of the reaction on an automated Edman sequencer, performing 9 cycles. If using a commercial facility, have them provide you with quantified data (in pmol) for all amino acid residues in all sequencing cycles.

4. Also subject 1 μL of the uncapped dodecamer library (X$_{12}$-amide) to Edman sequencing. To establish the relative quantities of each residue in the starting mixture, transfer the raw sequence data from the uncapped library to a spreadsheet. Let AVG_{Xxx} be equal to the average number of pmol across all nine sequencing cycles for a given amino acid (take the sum of the values in all cycles and divide by 9). Let AVG_{Tot} be equal to the overall average number of pmol per amino acid (take the sum of the AVG_{Xxx} values excluding AVG_{Trp} and divide by 18). The bias for a particular residue within the library is taken as the ratio AVG_{Xxx}/AVG_{Tot}.

5. Correct the data from the protease cleaved peptide library for bias in the starting mixture by dividing each quantity by the appropriate AVG_{Xxx}/AVG_{Tot} value. Normalize the data by dividing each data point by the average value found within a given sequencing cycle. The average value of the normalized data should then be 1, with positively selective residues having values greater than 1 and negatively selected residues having values less than 1. Sample normalized data are shown in **Fig. 2a**.

6. Repeat the Ac-X$_{12}$-NH$_2$ cleavage and sequencing, adjusting the amount of protease used to aim for 500 pmol total amino acid residue in the first sequencing cycle, assuming a linear relationship between enzyme concentration and product formation. If the protease does not appear able to cleave the fully degenerate mixture to a sufficient extent, adding more protease or using a partially degenerate primary library may be necessary (*see* **Note 5**).

3.3. Design and Preparation of Secondary and Tertiary Peptide Libraries

On the basis of the results from screening the primary peptide library Ac-X$_{12}$-NH$_2$, a secondary peptide library is designed in which the residues most highly selected downstream of the cleavage site (generally at the P1′ through P4′ positions) are fixed. Random degenerate sequence is incorporated at typically five positions upstream of the cleavage site. Additional amino acid residues are incorporated at flanking positions: at least one position at the very amino terminus of the peptide should be fixed, and a biotinyl-lysine residue is placed at the very carboxy-terminus (an alternative is discussed in **Note 6**). "Long chain" biotin, which incorporates an aminohexanoic acid spacer in between the biotin moiety and the lysine sidechain, should be used. As an example, M-A-X-X-X-X-X-L-R-G-A-A-R-E-K(biotin)-amide was used to profile matrix metalloproteinases *(6)*. Ideally this peptide would be cleaved exclusively at the desired bond. In practice such precisely directed cleavage is not always achieved with a secondary library, and often it is necessary to subsequently design a tertiary library that incorporates at least one fixed residue upstream of the cleavage site (for example M-G-X-X-P-X-X-L-R-G-G-G-E-E-K(biotin) for matrix metalloproteinases; also *see* **Note 7**).

The method depends on complete removal of uncleaved peptides and carboxy-terminal fragments using immobilized avidin prior to Edman sequencing. For reasons not totally clear, crude synthetic peptides with C-terminal biotinyl-lysine residues are not stiochiometrically biotinylated. Accordingly, secondary and tertiary libraries must be purified on a monomeric (reversible binding) avidin column before use.

1. Prepare a 5 mL monomeric avidin column by passing the following solutions over the column by gravity flow:

    a) 3×8 mL 2 mM biotin in PBS

    b) 50 mL 0.1 M glycine, pH 2.8

    c) 4×8 mL PBS

Make sure that the pH of the runoff from the column is close to neutral.

2. Dilute 150 µL of 4 mg/mL secondary library with 2 mL PBS and load onto the column. Wash with:

    a) 4×0.5 mL, then 1×9 mL PBS

    b) 40 mL 50 mM NH_4OAc

3. Elute bound peptides with 40 mL 0.2 M HOAc. Recover column by washing with PBS until the flowthrough has a pH value around neutral. The column may be reused about five times. Store the column at 4°C after equilibrating to PBS containing 0.02% sodium azide.

4. Freeze eluate in a dry ice/ethanol bath and lyophilize for 2 days. Resuspend residue in 150 µL of 25 mM phosphate or 25 mM HEPES, pH 7.4, as appropriate. Store frozen at −20°C.

3.4. Analysis of Secondary and Tertiary Peptide Libraries

1. Prepare the peptide cleavage mixture:

    14 µL avidin-purified secondary library

    2 µL ddH$_2$O

    2 µL 10× reaction buffer

2 µL 10× protease (diluted in reaction buffer)

For the initial run, use the same concentration of protease that was used for cleavage of the primary peptide library.

2. Incubate at 37°C for 2 h, and then quench by heating to 100°C for 2 min.

3. During the incubation, wash 400 µL of avidin agarose at least five times with 1 mL 50 mM NH_4HCO_3, 5 min per wash. Pellet beads at 4,000 rpm for 2 min in a microcentrifuge between washes.

4. Suspend the washed avidin agarose in an equal volume of 50 mM NH_4HCO_3 and add to the quenched protease reaction. Rotate the tube at room temperature for 1–2 h. Transfer the suspension to a disposable column. Collect the flowthrough and combine with 5 × 200 µL 50 mM NH_4HCO_3 washes. Dry the fractions in a speedvac, resuspend each pellet in 200 µL ddH_2O and combine into a single tube. Dry the combined fractions again in a speedvac. Resuspend the residue in 20 µL ddH_2O and analyze the entire sample by automated Edman degradation. Perform a number of cycles sufficient to sequence one cycle past the one corresponding to the P1 position (*see* **Note 8**).

5. Also sequence 2 µL of the undigested peptide library to determine bias in the starting mixture. In a spreadsheet, calculate AVG_n for each degenerate position by adding the number of pmol of all residues except Trp for that position and dividing by 18. The bias for each residue at each position, $BIAS_{Xxx,n}$ is calculated by dividing the pmol quantity by AVG_n.

6. To correct the cleaved peptide data for bias in the starting mixture, divide each pmol quantity by its respective $BIAS_{Xxx,n}$ value. Normalize the data by dividing each corrected pmol value by the average pmol value within the same sequencing cycle. Sample normalized data are shown in **Fig. 2b**.

7. Repeat the digestion, avidin treatment and Edman sequencing, adjusting the amount of protease to provide 500 pmol in the first sequencing cycle. If the amount of cleavage is too low, or if the extent of misdirected cleavage is high, the library may need to be redesigned (*see* **Note 9**).

4. Notes

1. By convention the positions surrounding a protease cleavage site are defined as Pn...P3-P2-P1-P1′-P2′-P3′...Pn′, where cleavage occurs between the P1 and P1′ positions.

2. Most commercial facilities are capable of generating partially or fully degenerate peptide libraries. Mixture positions can

either be made using a split-and-mix protocol *(20)* or using isokinetic mixtures *(21)* with comparable results. Because neither method provides a perfectly equimolar mixture of all amino acids, it is necessary to correct data from the peptide library screens for bias in the amino acid distribution. The actual distribution of residues is determined by sequencing an uncapped version of the library (described in **Subheading 3.2** above). The uncapped and acetylated versions of the library should be prepared by splitting the resin at the end of synthesis into two parts, one that will be acetylated and one left with a free amino terminus. Likewise, the biotinylated secondary and tertiary peptide libraries should also be made in capped and uncapped versions, with the acetylated peptides used to establish the extent of directed cleavage (*see* **Note 6**). Cysteine is left out of all mixture positions due to difficulties with oxidation. Though methionine also has a tendency to oxidize, it is included in the mixtures since data are normalized to the actual methionine content of the library. Tryptophan suffers from poor yield in Edman sequencing, generally precluding its analysis, but is included in the mixtures in the event that it is required at some position by the protease. Peptide mixtures cannot be purified by HPLC prior to use. Crude peptides carry a significant amount of trifluoroacetic acid with them that is left over from synthesis. It is important to check the pH of any reaction mixtures containing peptide libraries to ensure that the buffering capacity is not overwhelmed and to increase the buffer concentration as needed.

3. A source of highly purified, active protease is an important prerequisite for peptide library screening. All expression systems have the potential for contaminating protease activities to copurify with the protease of interest. In some cases additional purification steps will be required subsequent to a standard affinity purification procedure. It is strongly recommended that a protease inactive point mutant be prepared using the same expression system as the wild type enzyme. The mutant should be used in a mock digest of $Ac\text{-}X_{12}\text{-}NH_2$ to ensure that no signal on the sequencer arises from contaminating protease activities.

4. Generally one should use a reaction buffer that is optimized for the protease of interest. However, some common buffers are not compatible with peptide sequencing and should be avoided. Primary amine containing buffers such as Tris react with sequencing reagents and thus cannot be used. The common Tris substitute buffer HEPES, which has tertiary amino groups, produces an artifact peak on the sequencing chromatogram that obscures glutamine and threonine in later sequencing cycles. If possible, non amine-based buffers such as phosphate or acetate should be used.

5. There are a number of potential reasons why insufficient cleavage of the primary library is observed. Simply increasing the amount of protease used may solve the problem. However, the protease concentration should be kept below 1 µM if possible, as the protease itself will contribute to the signal on the Edman sequencer. If it becomes necessary to use higher levels, either removing the protease subsequent to digestion with an affinity tag, or running a mock sample with protease alone should be done to ensure that any signal observed on the sequencer arises from the peptide library and not the protease. A common reason for observing an insufficient degree of cleavage is that the enzyme is not fully in its active form. In such cases, activation conditions for the protease must be found to use the method. If a protein substrate is known, it is sometimes possible to compensate for low activity by generating a partially degenerate primary library in which amino terminal residues are fixed. For example, the peptide library acetyl-K-K-K-P-T-P-X-X-X-X-X-A-K-amide, based on the cleavage site from a known protein substrate, was used to determine the cleavage motif for anthrax lethal factor, which did not efficiently cleave a random dodecamer *(18)*.

6. The use of a high molecular weight branched dendrimer bearing multiple copies of the secondary library has been used as an alternative a biotin tag *(15)*. In this scheme, the amino-terminal cleavage products are separated from unreacted peptides and carboxy-terminal fragments by size exclusion chromatography. This modification eliminates the requirement of prepurifying the secondary library on monomeric avidin, and appears to give rise to a lower background peptide signal compared with the use of biotinylated peptides.

7. A tertiary library is necessary if more than 10% of the peptides in the secondary library are cleaved at a bond other than the intended one. The most straightforward way to estimate the extent of misdirected cleavage is to digest an amino-terminally acetylated version of the secondary library alongside the unblocked library. The digest is subjected to Edman sequencing directly without immobilized avidin treatment. Most of the amino acid in the first sequencing cycle should correspond to the intended P1′ residue in the library, and the amount of misdirected cleavage can be calculated from the quantity of degenerate sequence in the first few cycles.

8. Sequencing into the P1′ residue indicates the quantity of residual uncleaved peptide that did not bind to the avidin column. This amount should be low (5% of the total peptide) in order for the signal in the degenerate positions to be above background. A high level or residual uncleaved peptide could mean that insufficient avidin agarose was used to remove the

biotinylated peptides after digestion. Alternatively, it could indicate the presence of non-biotinylated peptides, either due to insufficient purification on the monomeric avidin column or to misdirected cleavage in the carboxy-terminal fixed positions (discussed in **Note 9**).

9. A high degree (>30%) of misdirected cleavage creates difficulties in data interpretation, as many of the cleaved peptides will be out of register (falling within an incorrect sequencing cycle), producing misleading data. Misdirected cleavage carboxy-terminal to the cleavage site will result in high background at the degenerate positions. There are a number of guidelines for designing secondary and tertiary libraries that can help avoid misdirected cleavage. Having a relatively small number of residues (five or less) amino-terminal to the cleavage site can help. This provides the protease with fewer places to cleave the peptide, and most proteases require three or four residues amino-terminal to the cleavage site. On the carboxy-terminal side, incorporating residues at P3′ and P4′ that are *deselected* by the protease at P1′ and P2′ will help avoid unwanted downstream cleavage. In addition, incorporating d-amino acid residues at P5′ and beyond helps restrict cleavage to the center of the peptide, since these are generally not allowed close to the cleavage site.

Acknowledgments

Synthesis of all peptide libraries described in the text was done at the Tufts University Medical School Core Facility by Michael Berne. I gratefully acknowledge Beth Piro for peptide sequencing and Lewis Cantley for mentoring and support during development of the methods described here.

References

1. Lopez-Otin, C. and Overall, C. M. (2002) Protease degradomics: a new challenge for proteomics. *Nat. Rev. Mol. Cell. Biol.* 3, 509–519.

2. Rawlings, N. D., Tolle, D. P., and Barrett, A. J. (2004) MEROPS: the peptidase database. *Nucleic Acids Res.* 32, D160–D164.

3. Mahajan, N. P., Harrison-Shostak, D. C., Michaux, J., Herman, B. (1999) Novel mutant green fluorescent protein protease substrates reveal the activation of specific caspases during apoptosis. *Chem. Biol.* 6, 401–409.

4. Vanderklish, P. W., Krushel, L. A., Holst, B. H., Gally, J. A., Crossin, K. L., Edelman, G. M. (2000) Marking synaptic activity in dendritic spines with a calpain substrate exhibiting fluorescence resonance energy transfer. *Proc. Natl. Acad. Sci. USA* 97, 2253–2258.

5. Wei, Q., Seward, G. K., Hill, P. A., et al. (2006) Designing ^{129}Xe NMR biosensors for matrix metalloproteinase detection. *J. Am. Chem. Soc.* 128, 13274–13283.

6. Turk, B. E., Huang, L. L., Piro, E. T., and Cantley, L. C. (2001) Determination of protease

cleavage site motifs using mixture-based oriented peptide libraries. *Nat. Biotechnol.* 19, 661–667.

7. Boyd, S. E., Pike, R. N., Rudy, G. B., Whisstock, J. C., and Garcia de la Banda, M. (2005) PoPS: a computational tool for modeling and predicting protease specificity. *J. Bioinform. Comput. Biol.* 3, 551–585.

8. Meldal, M., Svendsen, I., Breddam, K., and Auzanneau, F. I. (1994) Portion-mixing peptide libraries of quenched fluorogenic substrates for complete subsite mapping of endoprotease specificity. *Proc. Natl. Acad. Sci. USA* 91, 3314–3318.

9. Backes, B. J., Harris, J. L., Leonetti, F., Craik, C. S., and Ellman, J. A. (2000) Synthesis of positional-scanning libraries of fluorogenic peptide substrates to define the extended substrate specificity of plasmin and thrombin. *Nat. Biotechnol.* 18, 187–193.

10. Matthews, D. J. and Wells, J. A. (1993) Substrate phage: selection of protease substrates by monovalent phage display. *Science* 260, 1113–1117.

11. Salisbury, C. M., Maly, D. J., and Ellman, J. A. (2002) Peptide microarrays for the determination of protease substrate specificity. *J. Am. Chem. Soc.* 124, 14868–14870.

12. Dekker, N., Cox, R. C., Kramer, R. A., and Egmond, M. R. (2001) Substrate specificity of the integral membrane protease OmpT determined by spatially addressed peptide libraries. *Biochemistry* 40, 1694–1701.

13. Turk, B. E., Cantley, L. C. (2004) Using peptide libraries to identify optimal cleavage motifs for proteolytic enzymes. *Methods* 32, 398–405.

14. Martins, L. M., Turk, B. E., Cowling, V., et al. (2003) Binding specificity and regulation of the serine protease and PDZ domains of HtrA2/Omi. *J. Biol. Chem.* 278, 49417–49427.

15. Cuerrier, D., Moldoveanu, T., and Davies, P. L. (2005) Determination of peptide substrate specificity for μ-calpain by a peptide library-based approach: the importance of primed side interactions. *J. Biol. Chem.* 280, 40632–40641.

16. Murata, C. E., Goldberg, D. E. (2003) Plasmodium falciparum falcilysin: a metalloprotease with dual specificity. *J. Biol. Chem.* 278, 38022–38028.

17. Turk, B. E., Lee, D. H., Yamakoshi, Y., et al. (2006) MMP-20 is predominately a tooth-specific enzyme with a deep catalytic pocket that hydrolyzes type V collagen. *Biochemistry* 45, 3863–3874.

18. Turk, B. E., Wong, T. Y., Schwarzenbacher, R., et al. (2004) The structural basis for substrate and inhibitor selectivity of the anthrax lethal factor. *Nat. Struct. Mol. Biol.* 11, 60–66.

19. Udenfriend, S., Stein, S., Bohlen, P., Dairman, W., Leimgruber, W., and Weigele, M. (1972) Fluorescamine: a reagent for assay of amino acids, peptides, proteins, and primary amines in the picomole range. *Science* 178, 871–872.

20. Lam, K. S., Lake, D., Salmon, S. E., et al. (1996) A one-bead one-peptide combinatorial library method for B-cell epitope mapping. *Methods* 9, 482–493.

21. Ostresh, J. M., Winkle, J. H., Hamashin, V. T., and Houghten, R. A. (1994) Peptide libraries: determination of relative reaction rates of protected amino acids in competitive couplings. *Biopolymers* 34, 1681–1689.

Chapter 6

High Throughput Substrate Phage Display for Protease Profiling

Boris Ratnikov, Piotr Cieplak, and Jeffrey W. Smith

Summary

The interplay between a protease and its substrates is controlled at many different levels, including coexpression, colocalization, binding driven by ancillary contacts, and the presence of natural inhibitors. Here we focus on the most basic parameter that guides substrate recognition by a protease, the recognition specificity at the catalytic cleft. An understanding of this substrate specificity can be used to predict the putative substrates of a protease, to design protease activated imaging agents, and to initiate the design of active site inhibitors. Our group has characterized protease specificities of several matrix metalloproteinases using substrate phage display. Recently, we have adapted this method to a semiautomated platform that includes several high-throughput steps. The semiautomated platform allows one to obtain an order of magnitude more data, thus permitting precise comparisons among related proteases to define their functional distinctions.

Key words: Substrate phage display, Substrate, Protease, Specificity, Proteolysis, Filamentous phage, M13 coat protein, 3 gene protein.

1. Introduction

Sequence comparisons from over 100 genomes shows that peptidases comprise about 2% of all gene products *(1)*, and 50 years of research show that proteolysis plays an important role in most biological processes. Yet, we still have little knowledge on which protein substrates are cleaved by a particular protease, and we lack the technology for localizing and quantifying proteolytic events in cells and whole animals. One strategy for addressing these issues is to define the fine substrate recognition profile of individual proteases and then use this information to predict

Toni M. Antalis and Thomas H. Bugge (eds.), *Methods in Molecular Biology, Proteases and Cancer, Vol. 539*
© Humana Press, a part of Springer Science+Business Media, LLC 2009
DOI: 10.1007/978-1-60327-003-8_6

physiologic substrates and to design imaging agents and other probes of that particular protease.

Defining the substrate specificity of a protease involves assignment of amino acid preferences for each of the subsites within the catalytic cleft. These sites are normally referred to as primed (on the right side of the scissile bond) and unprimed (on the left side of the scissile bond) according to the definitions of Berger and Schechter (2). Different approaches have been developed for acquiring this information (for review see **ref.**3). These include the use of peptide substrate libraries, as well as biological display using bacteriophage and E. coli. Filamentous phage display (4) was first used for protease specificity profiling by Matthews and Wells in 1993 (5). Since then, substrate phage display has been used to profile a large number of individual proteases (6–11).

Substrate phage display is typically carried out by displaying a randomized peptide substrate as a fusion protein with the gene 3 protein (g3p) of filamentous M13 bacteriophage. Since g3p is expressed in 3–5 copies per phage, there is a polyvalent display of the substrate peptide. This peptide is flanked on its C-terminal side by g3p and a "spacer" designed to keep the randomized sequence in a disordered conformation. An affinity tag is normally appended to the N-terminal side of the peptide, and this is used to separate cleaved from uncleaved phages during the selection process (**Fig. 1**). Our group uses a library composed of randomized hexapeptides displayed on g3p and flanked by a FLAG epitope engineered at the NH$_2$ terminus of the g3p (12). The substrate phage library was generated using a modified version of the fUSE5 phagemid.

One of the major limitations of substrate phage display is the length of time required to obtain data that yield a comprehensive substrate recognition profile for a given protease. Recently we have incorporated several automated steps into the substrate phage method that allow substantially higher throughput, and that yield an order of magnitude more data. With this platform in place we believe it possible to define the substrate recognition specificity of every endopeptidase in the human genome within 5–10 years. Here we describe the detailed method linked to the semiautomated platform.

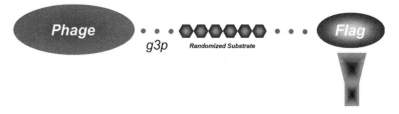

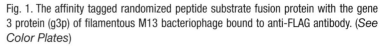

Fig. 1. The affinity tagged randomized peptide substrate fusion protein with the gene 3 protein (g3p) of filamentous M13 bacteriophage bound to anti-FLAG antibody. (*See Color Plates*)

2. Materials

2.1. Phage Propagation and Purification

1. NZY Broth: 5 g NaCl, 2 g MgSO$_4$·7H$_2$O, 5 g Bacto-yeast extract, NZ amine (casein hydrolysate), ddH$_2$O to 1 L. Adjust pH to 7.5 with NaOH and autoclave.

2. 100 mg/mL Kanamycin Sulfate in water, store at –20°C.

3. 20 mg/mL Tetracycline in 50% glycerol store at –20°C.

4. PEG/NaCl: 100 g PEG 8000 (Sigma catalog # P4463, 116.9 g NaCl, 475 mL water; stir until solutes dissolve (may be necessary to heat to 65° briefly to dissolve the last crystals of PEG). Sterile filter through a 0.22 μm membrane. Store at 4°C (total volume 600 mL).

5. K91Kan *E. coli* cells.

6. Glycerol stock solution 2×: 25 mM Tris, pH 8.0, 0.1 M MgSO$_4$, 65% Glycerol.

2.2. Phage Substrate Selection

1. Substrate phage library based on the fUSE5 phagemid with a FLAG epitope engineered at the NH$_2$ terminus of the gene III protein.

2. Stock solution of the protease of interest preferably at high concentration.

3. High affinity inhibitor for active site titration and protease inhibition.

4. Monoclonal anti-FLAG antibodies M2 (Sigma catalog # F3165).

5. Monoclonal anti-M13 antibodies (GE Healthcare catalog # 27-9421-01).

6. Dynabeads M-450 epoxy (Invitorgen catalog # 14011).

7. Dynal MPC-S magnetic separator (Invitrogen catalog # 120-20D).

8. Monoclonal anti-M13 antibodies HRP conjugated (GE Healthcare catalog # 27-9420-01).

9. OPD HRP substrate (Sigma catalog # P3804-100TAB).

10. HRP substrate buffer: 2.43 mL of 0.1 M Citric Acid, 2.57 mL of 0.2 M Na$_2$HPO$_4$, 5 mL of DI water.

11. 30% H$_2$O$_2$.

12. 4 M H$_2$SO$_4$.

13. 3 M (NH$_4$)$_2$SO$_4$.

14. PBS: 50 mM sodium phosphate, pH 7.4, 150 mM NaCl.

15. 100 mM sodium phosphate, pH 7.0.

16. TBS: 50 mM Tris, pH 7.2, 150 mM NaCl.

17. TBS-T: TBS, 0.1% Tween 20.

18. Polystyrene 96-well microplates (Costar catalog # 9018).

19. 50 mL centrifuge tubes (Corning catalog # 430828 or similar).

20. 100 mg/mL BSA (bovine serum albumin, Sigma catalog # A7906) in TBS.

21. Microplate reader.

22. Multichannel micropipettor.

23. PC with Graph Pad Prism or other data fitting software.

2.3. Substrate
Identification

1. K-91Kan *E. coli* glycerol stock.

2. Terrific Broth: 12 g bacto-tryptone, 24 g yeast extract, 4 mL (5.04) g glycerol, DI water 900 mL. Autoclave 90 mL portions in 125 mL polypropylene bottles. When cooled, to each bottle add 10 mL of separately autoclaved potassium phosphate buffer (0.17 M KH_2PO_4, 0.72 M K_2HPO_4).

3. LB Agar, 100 µg/mL Kanamycin: 10 g Tryptone, 5 g yeast extract, 5 g NaCl, 200 µL NaOH, 15 g Bacto Agar DI water to 1 L. Autoclave, cool to 50°C, add 1 mL 100 mg/mL Kanamycin – pour 100 mm Petri dishes at 20 mL/dish. Store at 4°C after agar solidifies.

4. NZY Agar, 100 µg/mL Kanamycin, 40 µg/mL Tetracycline: 5 g NaCl, 2 g $MgSO_4 \cdot 7H_2O$, 5 g Bacto-yeast extract, NZ amine (casein hydrolysate), 15 g Bacto Agar, DI H_2O to 1 L. Adjust pH to 7.5 with NaOH. Autoclave, cool to 50°C, add 1 mL 100 mg/mL Kanamycin – pour into Genetix QTray vented (Genetix catalog # 6023) at 200 mL/tray for automated colony picking using Genetix QPix colony picker or at 20 mL per 100 mm Petri dish. Store at 4°C after agar solidifies.

5. Multichannel pipettor.

6. Hamilton LabStar liquid handling robot for HTS phage diplay.

7. 96 deep well plate (Simport Bioblock deep well plate, catalog # T110-10) for HTS phage display or 15 mL sterile disposable round bottom culture tubes for low throughput phage display.

8. Genetix 96-well flat bottom1/2 height plate (Genetix catalog # X6011) for HTS phage display or sterile 1.5 mL Eppendorf centrifuge tubes for low throughput phage dispay.

9. Genetix BreatheSeal Film (catalog # E1005).

10. Aluminum sealing tape, (ISC BioExpress catalog # T-2420-2).

11. Glycerol stock solution.

12. Monoclonal anti-M13 antibodies.

13. Monoclonal anti-FLAG M2 antibodies peroxidase conjugated (Sigma catalog # A8592).

14. TBS-T: TBS, 0.1% Tween 20.

15. Polystyrene 96-well microplates.

16. 100 mg/mL BSA in TBS.

17. HRP Substrate Buffer.

18. OPD HRP substrate.

19. 30% H_2O_2.

20. Microplate reader.

21. ATR Multitron Incubator shaker.

2.4. Substrate Sequencing

1. FUSE5 forward primer: 5′–TAA TAC GAC TCA CTA TAG GGC AAG CTG ATA AAC CGA TAC AAT T–3′, 100 μM stock.

2. FUSE5 Super Reverse primer: 5′–CCG TAA CAC TGA GTT TCG TC–3′, 100 μM stock.

3. Platinum PCR Super Mix (Invitrogen catalog # 11306-016).

4. 96-well PCR plate (Abgene catalog # AB1000).

5. PCR thermocycler.

6. 50× TAE Buffer.

   a. 242 g Tris base

   b. 57.1 mL acetic acid

   c. 100 mL 0.5 M EDTA

   d. Add ddH$_2$O to 1 L and adjust pH to 8.5

7. 10 mg/mL ethidium bromide (Sigma catalog # E1510).

8. 1% agarose in TAE buffer, 0.5 μg/mL ethidium bromide.

2.5. Identification of Scissile Bonds

1. NZY Broth.

2. 96 deep well plate (Simport Bioblock deep well plate, catalog # T110-10) for HTS phage display or 15 mL sterile disposable round bottom culture tubes for low throughput phage display.

3. Costar Assay Block 1 mL plate (catalog # 3958).

4. Costar Storage Mat III (catalog # 3080).

5. Monoclonal anti-FLAG antibodies M2.

6. Monoclonal anti-M13 antibodies.

7. Dynabeads M-450 epoxy.

8. Dynabeads M-280 Tosylactivated (Invitrogen catalog # 14204).

9. 2,5-Dihydroxybenzoic acid, 99% (Aldrich catalog # 14,935-7).

10. Trifluoroacetic acid (Sigma Aldrich catalog # T6508).

11. MALDI AnchorChip target (Bruker Daltonics catalog # 209515) or other MALDI target.

12. 3 M $(NH_4)2SO_4$.

13. 100 mM sodium phosphate, pH 7.0.

14. TBS-T: TBS, 0.1% Tween 20.

15. V-bottom polypropylene plates (Costar catalog # 3357).

3. Methods

Prior to delving into procedures for substrate phage display, it is important to mention a few factors and pitfalls that have a significant impact on the success of the procedure. A major factor in the success of substrate phage display is the availability of an ample supply of the test protease, and a reliable method of quantifying its activity. Each time a protease is used for substrate phage its activity should be titrated with an active site inhibitor *(13)* whenever possible. When the test proteases are expressed and purified in our laboratories, more than 70% of the purified protein is active. With this level of activity, 500 μg to 1 mg of total protease are usually sufficient to complete the phage procedure. We are reluctant to move forward with phage selections for any protease in which less than 30–40% of the total protein has activity. With this quantitative information in hand, we have the basis for quantitative analysis and comparison of any of the data obtained in the course of the experiment.

Another factor that has a major impact on success is the quality of the K91Kan *E. coli* used for phage infection and propagation. These bacteria must be fresh; generated and used within 24 h. This is a very important requirement and should be followed strictly to obtain consistent results.

An important pitfall in this procedure, as it is with any phage display project, is the possibility of phage contamination. There are really two different manifestations of this problem. The first is the simple contamination of laboratory tools and instruments with a single phage, which then begins to dominate all successive phage amplifications and overrides any selection process. This can be prevented by highly stringent sterile techniques, the use of pipette tips containing filters, and frequent washing of instrumentation. All buffers and solutions should be sterilized either by autoclaving or 0.2 μm filtration.

In another manifestation this problem arises as the propagation of a phage lacking the affinity tag, which then appears as "cleaved" even when it is not *(8)*. This is not a problem for the first round of selection, but can be an issue after the second round, when some phages have adapted to the selection pressure by eliminating the tag from their sequence. To overcome this issue, we include an affinity isolation of the tag-bearing phage prior to the next round of selection.

A protease phage display project involves four basic steps: (1) substrate selection, (2) substrate identification, (3) substrate sequencing, and (4) scissile bond assignment. Steps 2–4 have been automated by our group. The procedure for each step is described in detail later.

3.1. Phage Substrate Selection

These instructions assume that a FLAG-tagged phage library is available for substrate selection.

1. Before starting substrate selection, prepare M-450 M2 beads. Wash 5×10^8 M-450 epoxy magnetic beads three times by 1 mL 100 mM sodium phosphate buffer, pH 7.0 by magnetic separation. Resuspend the beads in 500 μL 0.5 mg/mL M2 antibody and add 250 μL 3 M $(NH_4)_2SO_4$. Incubate with end-over-end rotation or vortexing overnight at room temperature, making sure the beads are in suspension. In the morning collect the supernatant and measure light absorbance at 280 nm by spectrophotometry. Determine the efficiency of coupling by dividing by the absorbance of the starting antibody solution adjusted by dilution factor of 1.5. Usually 75–90% of the antibody is coupled under these conditions. Resuspend the beads in 1 mL of 1 mg/mL BSA in PBS and incubate with end-over-end rotation or vortexing for 3 h at room temperature. Store the beads at 4°C.

2. The day of substrate selection, perform active site titration of the protease or specific activity determination by some other method to be able to reproduce the conditions of the experiment.

3. For proteases requiring extreme pH for optimal activity, determine neutralization conditions.

4. Determine the concentration of the phage particles in the stock solution of the phage library. Dilute the phage in TBS to a concentration approximately 10^{12} particles/mL. Scan the diluted sample from 240 to 320 nm in a UV spectrophotometer. There should be a broad peak from 260 to 280 nm with a slight maximum at 269 nm. Measure A_{269} and A_{320} and calculate the net A_{269} by subtracting A_{320} from A_{269}. Calculate the virion concentration in particles per mL based on the following

formula: phage particles/mL = net $A_{269} \times 6 \times 10^{16}/N$, where N = the number of nucleotides/phage genome. The genome size of a fUSE5 phage = 9,206 bases.

5. Add 1.2×10^{12} phage to the final volume of 1.2 mL in protease reaction buffer. Take out two 0.4 mL aliquots for negative control and protease treatment. To the first 0.4 mL aliquot add the protease in the protease buffer, to the final concentration of 0.2 μM of active enzyme. To the second and third aliquots add the same volume of the protease buffer. Incubate the mixtures for 2 h at the optimal temperature for the protease. Upon completion of the incubation add protease inhibitor to all the samples and adjust the pH to 7.2–7.4. Bring the volumes up to 1 mL with TBST.

6. During the incubation wash 0.8 mL worth of M-450 M2 bead suspension in 1 mL TBST three times using magnetic separator. After the second wash, split the magnetic bead suspension in four 0.25 mL aliquots and remove the supernatant after magnetic separation from two of them. Resuspend the beads with protease treated and 1 aliquot of control phage in one tube each and incubate for 1 h at ambient temperature with rotation end-over-end. The second control aliquot of phage will be used as the before-depletion control. Do not use vortexing to keep the beads in suspension when doing phage separations. After the incubation is over, remove the supernatant from the remaining two tubes of M-450 M2 beads by magnetic separation. Collect the unbound phage from the control and the protease treated phage by magnetic separation and mix with the fresh M-450 M2 beads from the remaining tubes. Incubate at ambient temperature with end-over-end rotation for 3 h. By the end of incubation, collect the supernatants by magnetic separation and determine the efficiency of depletion the following day by ELISA. Store the phage samples at –70°C.

7. Coat a 96-well microtiter plate with 100 μL/well of 5 μg/mL anti-M13 antibody in TBS at 4°C overnight.

8. The same night streak an LB agar 100 μg/mL Kanamycin Petri dish with K91Kan *E. coli* cells and incubate at 37°C overnight. Remove the dish from the incubator in the morning of the following day and place at 4°C until use.

9. In the morning of the next day block the plate with 120 μL/well of 100 mg/mL BSA in TBS for 1 h at 37°C.

10. Prepare seven dilutions of phage with known concentration (standard) from 10^{11} phage/mL with a step of 1:2 by combining 0.35 mL of each dilution with 0.35 mL TBST. Dilute the before-depletion control, the protease treated and the control M2 depleted phage 1:20 by adding 35 μL of the solution to

665 μL TBST. Serially dilute the phage three times by combining 0.35 mL with 0.35 mL TBST. Add 100 μL/well from each dilution in triplicate. Incubate for 1 h at 37°C. When performing ELISA for subsequent rounds of selection, it is important to use phage from the starting stock used for this round of selection as standard, to obtain consistent results.

11. After incubation, wash the plate with 120 μL/well TBST three times. Add 100 μL/well anti-M13 monoclonal antibody – HRP conjugate diluted in TBST as recommended by the manufacturer. Incubate for 1 h at 37°C.

12. Upon completion of the incubation prepare the HRP substrate solution in the HRP substrate buffer. Wash the plate three times with 120 μL/well TBST. Add 4 μL/10 mL of 30% H_2O_2 to the substrate solution and apply at 100 μL/well to the plate. Incubate with shaking until color develops and add 50 μL 4 M H_2SO_4 to stop the reaction. Read the plate in a microtiter plate reader at 490 nm.

13. Plot the A_{490} vs. phage concentration for the standard curve. Fit the data points with the two site binding equation:
$A_{490} = B_{max_1} C/(K_{d_1} + C) + B_{max_2} C/(K_{d_2} + C)$, where B_{max_1} and B_{max_2} are the maximum binding per site, and K_{d_1} and K_{d_2} are respective dissociation constants, and C is phage concentration. Apparently, the anti-M13 antibody recognizes two different classes of binding sites on the M13 coat protein expressed by the phage. Determine the phage concentration for the samples from the standard curve. Typically 98–99% of phages are depleted in the untreated sample as observed for the first round of selection, while the difference in depletion between the protease treated and untreated phage represents the enrichment for protease substrates (**Fig. 2**).

14. Start an overnight culture of K91Kan *E. coli* by inoculating 3 mL of LB 100 μg/mL Kanamycin with a single colony of a K91Kan cells streaked the night before. Incubate at 37°C overnight with rotation at 250 rpm.

15. Take 100 μL of the overnight culture and add to 9.9 mL Terrific Broth in a sterile 50 mL centrifuge tube. Incubate at 37°C with shaking at 250 rpm for 2 h. After 2 h start monitoring A_{600} of the culture by taking out 0.1 mL and adding to 0.9 mL of DI water and measuring A_{600}. Multiply the obtained number by 10 to get the A_{600} for the culture.

16. When A_{600} = 2 AU (10^{10} cells/mL), slow down the shaking to 50 rpm for 5 min. Add an appropriate amount of the bacterial culture and the correspondingly appropriate volume of Terrific Broth so that after the addition of the substrate enriched phage, both the *E. coli* and the phage will be at 2 × 10^9 particles/mL. Shake the culture at 100 rpm at 37°C

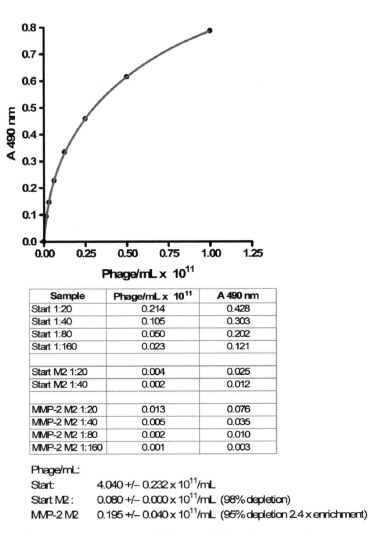

Sample	Phage/mL x 10^{11}	A 490 nm
Start 1:20	0.214	0.428
Start 1:40	0.105	0.303
Start 1:80	0.050	0.202
Start 1:160	0.023	0.121
Start M2 1:20	0.004	0.025
Start M2 1:40	0.002	0.012
MMP-2 M2 1:20	0.013	0.076
MMP-2 M2 1:40	0.005	0.035
MMP-2 M2 1:80	0.002	0.010
MMP-2 M2 1:160	0.001	0.003

Phage/mL:
Start: $4.040 +/- 0.232 \times 10^{11}$/mL
Start M2 : $0.080 +/- 0.000 \times 10^{11}$/mL (98% depletion)
MMP-2 M2 $0.195 +/- 0.040 \times 10^{11}$/mL (95% depletion 2.4 x enrichment)

Fig. 2. A typical standard curve for an M13 phage ELISA and the results of the first round of selection of MMP-2 substrates.

for 15 min. Transfer the infected cells into 10× volume of NZY broth containing 0.22 μg/mL tetracycline. Shake the culture for 35 min at 250 rpm, 37°C.

17. Add 0.001× volume of 20 mg/mL Tetracycline to the culture and incubate overnight at 250 rpm 37°C.

18. Determine the number of infected cells. Add 7 μL of culture to 63 μL LB medium. Perform serial dilutions three times by adding 20 μL of bacteria from the previous dilution to 180 μL LB. Plate 100 μL of each dilution onto 100 μg/mL Kanamycin, 40 μg/mL Tetracycline NZY agar 10 cm Petri dish. Incubate overnight at 37°C. In the morning of the following day, count the colonies on each dish. Multiply by the appropriate dilution factor and then by 10 to get the

number of infected cells per mL. Knowing the total number of the infected cells is useful in estimating how many individual phage clones have been obtained during infection.

19. The following morning centrifuge the culture in a sterile centrifuge tube or bottle at 2,400 × g for 10 min at 4°C. Without disturbing the cell pellet, transfer the phage-containing supernatant into a clean sterile centrifuge container and centrifuge again at 6,200 × g for 10 min at 4°C. Carefully pour the supernatant into a clean sterile container with a screw cap.

20. Add 0.15× volume of PEG/NaCl solution to the phage, close the container and mix thoroughly by inverting the container. Incubate on ice for at least 4 h or overnight at 4°C. The medium should turn cloudy.

21. Pellet the phage at 6,200 × g for 40 min at 4°C. Carefully discard the supernatant without disturbing the pellet. Remove the residual supernatant by briefly centrifuging the pellets and carefully aspirating the rest of the liquid.

22. Add 5 mL of sterile TBS and shake the pellets at 150 rpm at 37°C for approximately 30 min to resuspend the pellet.

23. Centrifuge the phage solution at 10,100–22,700 × g for 10 min at 4°C. Transfer the supernatant into a fresh sterile centrifuge container.

24. Add 0.15× volume of PEG/NaCl solution to the container and mix thoroughly by inversion. Allow the phage to precipitate for 1 h on ice. A heavy precipitate should appear.

25. Centrifuge the phage at 10,100 × g for 40 min at 4°C. Remove the supernatant.

26. Add 2 mL of sterile TBS to the pellet and allow the pellet to soften for 1 h. Vortex the pellet again to bring the phage into solution

27. Clear the supernatant by centrifuging at 10,100–22,700 × g for 10 min at room temperature or 4°C.

28. Collect the supernatant. Determine the phage concentration as described in **step 4** above. Store the phage short term at 4°C, long term at –70°C. The first round of selection is complete.

29. Repeat the above steps for the second round. If more than two rounds of selection are required, as can be the case if poor enrichment for substrate phage is observed after the second round, we recommend enriching the phage from the second round for high FLAG expressers. The reason for this is that after each amplification, there is a loss of tag expressing phage and it can be quite substantial after the second round. We routinely enrich phage for FLAG expression before proceeding to substrate identification.

30. Take 200 µL M2 M450 beads (10^8) and wash three times with TBST by magnetic separation. Add 2.5×10^{12} phage to the beads and bring the volume up to 1 mL with TBST. Incubate overnight with end-over-end rotation.

31. The following day, make 10 mL of 0.005% Brij 35 solution in DI water and sterilize it by filtration through a 0.2 µm filter.

32. Prepare 0.05% TFA solution in 0.005% Brij 35 solution.

33. Wash the beads three times with 1 mL TBST by magnetic separation after discarding the supernatant followed by two-time wash with 0.005% Brij solution. Remove any traces of liquid from the beads and add 100 µL of 0.05% TFA to the beads. Shake for 10 min at room temperature making sure the beads stay in suspension. Collect the supernatant by magnetic separation and mix with 100 µL TBS. Determine the phage concentration as described in **step 4** above. Multiply the obtained number by the total volume to get the total number of phage. Typically we get around 5×10^{11} phage after enrichment.

3.2. Verification of Substrate Phage

After the final round of selection is complete, one needs to verify that individual phage clones are bearing substrates prior to nucleotide sequencing. This is done by substrate phage ELISA of the media from bacterial cultures grown from single colonies of K91Kan cells bearing a single phage clone. To ensure that we select phage clones with adequate FLAG epitope expression, we use a positive control phage with high FLAG expression level (same or higher than the starting library) and select only those substrate phage clones, whose FLAG expression level is above 0.5 of that of the control. This is done to minimize the error of measurement of the degree of hydrolysis of substrate phage, which is determined by comparing the amount of FLAG with and without exposure to protease. In addition, the level of expression of the FLAG epitope reflects the number of copies of the substrate peptide per phage, thus allowing us to select high expressers for MALDI TOF analysis used for identification of the scissile bond in the substrate peptide.

1. Perform enrichment of the phage from the last round of selection for FLAG epitope expression as described in **Subheading 3.1, steps 30–33**.

2. Prepare NZY agar QPix trays or 10 cm Petri dishes by pouring 200 or 20 mL of NZY agar with 100 µg/mL Kanamycin, 40 µg/mL Tetracycline per tray/dish for HTS or LTS versions of the procedure respectively. Store the NZY agar trays/dishes at 4°C in a plastic bag.

3. Start an overnight 3 mL culture in LB 100 μg/mL Kanamycin from a freshly streaked colony. Incubate overnight at 37°C, shaking at 250 rpm.

4. Next morning add 100 μL of the overnight culture to 10 mL of Terrific Broth. Incubate at 37°C, shaking at 250 rpm.

5. After 2 h of incubation start monitoring A_{600} of 1:10 diluted culture in DI water until it reaches 0.2 AU. Slow the shaking down to 50 rpm for 5 min.

6. Place the Qtrays or Petri Dishes in a 37°C incubator.

7. Add 1 mL of the K91Kan culture (10^{10} cells/mL at A_{600} = 2.0) to 9 mL Terrific Broth and supplement with 10^9 FLAG enriched phage from the last round of selection. Continue incubation at slow shaking for 15 min.

8. Prepare 10 mL NZY broth with 0.22 μg/mL tetracycline from 20 mg/mL stock. Add 1 mL of infected culture to 10 mL of NZY broth 0.22 μg/mL Tetracycline. Shake 35 min at 250 rpm, 37°C.

9. Add 10 μL of 20 mg/mL stock of Tetracycline to the culture.

10. Dilute the culture 50-fold in NZY Broth and plate 0.75 mL per QPix tray or 75 μL per 10 cm Petri dishes. Place the trays in a 37°C incubator with 70% humidity. The Petri dishes can go in a 37°C incubator without humidification.

11. The next morning pick colonies using the QPix robot in the case of QPis trays or manually from the Petri dishes into 96-well deep-well 2 mL plates filled with 1 mL per well of NZY broth supplemented with 20 μg/mL Tetracycline. After picking, put the BreatheSeal film on the plates and incubate in the ATR Multitron incubator shaker at 750 rpm, 37°C, 70% humidity. For a small scale project, a regular shaking incubator can be used at 250 rpm, 37°C.

12. The same day, coat 96 well ELISA plates with 100 μL/well of 5 μg/mL anti-M13 antibody in PBS at 4°C overnight.

13. In the morning, block the plates with 120 μL of 100 mg/mL BSA in PBS for 1 h at 37°C.

14. Remove the BSA and add 100 μL/well of the bacterial cultures from the deep well plates as shown in **Fig. 3**: To the BLNK wells add 100 μL NZY broth. To the + CTRL wells add 100 μL of 10^{12} phage/mL of the positive control clone in NZY broth.

15. Incubate the plates for 1 h at 37°C.

16. Wash the plates with 120 μL/well TBST four times and remove the residual buffer from the plate by gently tapping the plates face down against a paper towel.

17. Add 100 µL of (50 nM active enzyme) protease solution per well to the left half of the plate as shown in **Fig. 3**. To the right half add 100 µL/well of the protease buffer. Incubate for 2 h under the optimal conditions for the protease.

18. Wash the plates four times with 120 µL/well TBST and add 100 µL/well of HRP conjugated M2 antibody appropriately diluted in TBST. Incubate for 1 h at 37°C. Dissolve 1 tablet of OPD per 10 mL of HRP substrate buffer right before the end of the incubation period.

19. Wash the plates four times with 120 µL/well TBST, remove the residual buffer by gently tapping the plates face down against a paper towel. Add 4 µL of 30% H_2O_2 per 10 mL of OPD solution, mix thoroughly and dispense 100 µL/well. Shake the plates gently and observe color development. Usually the plates can be read 5 min after addition of the substrate, but one should exercise proper judgment to stop the reaction at an appropriate time, so as not to over or under develop the reaction. A target A_{490} value of 1 is optimal. Add 50 µL/well of 4 M H_2SO_4 to stop the reaction. Read the plates at 490 nm in a microtiter plate reader.

20. Analyze the data by first eliminating from consideration any clones with A_{490} less than 0.5 of that of the positive control. For the remaining clones calculate the degree of hydrolysis by dividing the A_{490} of the + protease well by that of the − protease well and subtracting the result from 1. Disregard

	1	2	3	4	5	6	7	8	9	10	11	12
A	BLNK	6	14	22	30	38	+ CTRL	6	14	22	30	38
B	BLNK	7	15	23	31	39	+ CTRL	7	15	23	31	39
C	BLNK	8	16	24	32	40	+ CTRL	8	16	24	32	40
D	1	9	17	25	33	41	1	9	17	25	33	41
			+ PROTEASE						− PROTEASE			
E	2	10	18	26	34	42	2	10	18	26	34	42
F	3	11	19	27	35	43	3	11	19	27	35	43
G	4	12	20	28	36	44	4	12	20	28	36	44
H	5	13	21	29	37	45	5	13	21	29	37	45

Fig. 3. Template for substrate phage ELISA. Bacterial cultures are applied to anti-M13 antibody coated 96-well plate at 100 µL/well. Each culture is applied to the plate twice: on the left half of the plate and symmetrically on the right half. The phage captured on the left half of the plate is treated with protease (*shaded area*), while their counterparts on the right are not. The degree of hydrolysis is determined by subtracting the ratios of A_{490} between the protease treated and nontreated samples from 1.

any clones with the degree of hydrolysis of less than 0.2. The substrate identification step is complete.

21. Prepare glycerol stocks of the substrate phage by mixing 100 μL of culture with 100 μL of glycerol stock solution in a Genetix 96-well flat bottom1/2 height plate. Store the glycerol stocks at −70°C. In some cases there is a pH requirement for proteolytic activity that is incompatible with phage capture by anti-M13 antibodies. In these cases, we perform protease treatment of PEG precipitated phage in the protease buffer with a pH necessary for optimal protease activity. After that we adjust the pH to neutral and perform the ELISA.

22. Prepare glycerol stocks of the substrate phage by mixing 100 μL of culture with 100 μL of glycerol stock solution in a Genetix 96-well flat bottom1/2 height plate. Seal the plates with aluminum sealing tape. Store the glycerol stocks at −70°C.

23. Centrifuge the remaining overnight cultures at 4,000 × g for 10 min at 4°C. Collect the supernatants into fresh deep 96-well plates. Add 0.15× volume of PEG/NaCl solution, cover the plate with a storage mat, mix by inversion end-over-end about 100 times, spin briefly at 300 × g and put at 4°C overnight. Centrifuge at 4,600 × g for 90 min at 4°C. Carefully discard the supernatant and remove residual liquid by gently tapping the plates face down against a paper towel.

24. Add 120 μL/well of the protease buffer and shake the plates at 750 rpm in a ATR Multitron shaker incubator or similar for 30 min.

25. Transfer 50 μL of the phage solution to 50 μL of protease buffer per well on the right side of a 96-well uncoated ELISA plate (**Fig. 3**). Transfer another 50 μL of the phage solution to 50 μL of 100 nM protease in protease buffer in the counterpart wells on the left side of the plate. Incubate at appropriate temperature for 2 h.

26. Neutralize the samples by addition of an appropriate amount of neutralization solution (determined earlier). Transfer the samples to an anti-M13 coated BSA blocked plate. Incubate for 1 h at 37°C.

27. Proceed as described in **steps 17–19** above.

3.3. Sequencing Phage Substrates

Once the protease substrate phage clones have been identified, the nucleotide sequence of the random insert is determined. This is done by PCR amplification of the region of the g3p bearing the randomized insert followed by sequencing of the PCR amplicons. The PCR procedure is performed in-house in a 96-well format and the sequencing of the resulting amplicons is outsourced.

1. Thaw the glycerol stocks of the substrate phage at room temperature. Cherry pick the substrate clones identified by substrate phage ELISA and inoculate 1,250 μL of LB 20 μg/mL Tetracycline per well of 2 mL 96-deep well plates, using QPix colony picking robot for an HTS phage display or a sterile Pasteur loop for an LTS one. Seal the plates with BreatheSeal film. Grow the inoculates overnight at 750 rpm, 37°C, 70% humidity in an ATR Multitron shaker incubator. For a small scale project, a regular shaking incubator can be used at 250 rpm, 37°C.

2. The following morning prepare the PCR master mix:

   a. 2 μL/sample 5 μM FUSE5 forward primer

   b. 2 μL/sample 5 μM FUSE5 Super Reverse primer

   c. 45 μL Platinum PCR Super Mix

3. Dispense 49 μL of the master mix per well of a 96-well PCR plate. Add 2 μL per well of the overnight culture. Run the PCR as follows:

   a. 72°C 10 min

   b. 94°C 3 min

   c. 34 × (94°C 50 s, 50°C 1 min, 72°C 1 min)

   d. 72°C 6 min

   e. Hold at 4°C.

4. To determine the quality of the PCR, run 10 PCR products per plate at 10 μL/lane on 1% agarose gel.

5. Freeze the PCR products and send out for sequencing.

3.4. Identification of Scissile Bonds

Once the sequences of phage substrates are known from nucleotide sequencing, the position of the scissile bond within the substrate peptide is determined directly from purified phage using MALDI TOF MS. We take advantage of the fact that the phage express sufficient levels of g3p, so that the peptide hydrolyzed from the phage by protease is present in adequate quantity to analyze by MS. Once the mass of the cleaved portion of the peptide substrate is known, this information can be combined with sequence information to pinpoint the position of the scissile bond.

1. Prepare the necessary amount of M-450 anti-M13 beads at 10^8 beads per sample by following the protocol described in **Subheading 3.1**. Use anti-M13 instead of M2 antibodies.

2. Prepare M2 coupled M-280 beads. Wash $1.29 × 10^9$ beads three times with 1 mL of 100 mM sodium phosphate pH 7.0 by magnetic separation. Resuspend the beads in 1 mL of 250 μg/mL M2 antibody in 100 mM sodium phosphate, pH 7.0.

3. Incubate with end-over-end rotation overnight at 37°C.

4. In the morning of the following day measure the A_{280} of the supernatant after magnetic separation. Determine the amount of antibody coupled. Typically 80–90% of antibody gets coupled to M-280 beads.

5. Remove the supernatant by magnetic separation and resuspend the beads in 1 mL of sterile 1 mg/mL BSA in TBS. Incubate with end-over-end rotation for 4 h at room temperature. Store at 4°C.

6. Thaw the glycerol stocks of substrate phage and inoculate 1.2 mL cultures of substrate phage in 2 mL 96-deep well plates and grow overnight as described in **Subheading 3.3** of **step 1**.

7. The following morning wash the M-450 anti-M13 beads three times with TBST. Resuspend in the original volume of TBST and dispense into an Assay Block 1 mL 96-well plate at 200 µL/well.

8. Spin down the cultures at 4,000 × g, 4°C for 10 min.

9. Remove TBST from the M-450 anti-M13 beads by magnetic separation and resuspend the beads in 1 mL of culture supernatant. Seal the plates with the storage mats and incubate with rotation end-over-end for 3 h at room temperature.

10. Wash the beads with 1 mL of the protease incubation buffer three times by magnetic separation and resuspend in 100 µL/well of 50–200 nM active protease. Incubate with vortex shaking overnight at the optimal temperature.

11. Alternatively, one can use PEG precipitation for buffer exchange, especially if the protease buffer has acidic or alkaline pH incompatible with the antibody capture of the phage. This approach provides good results as well, although there will be residual PEG left over. One has to test if PEG has an effect on the protease activity before using this method.

    a. Add 0.15× volume of PEG/NaCl solution, cover the plate with a storage mat, mix by inversion end-over-end about 100 times and incubate at 4°C overnight. Centrifuge at 4,600 × g for 90 min at 4°C. Carefully discard the supernatant and remove residual liquid by gently tapping the plates face down against a paper towel.

    b. Add 100 µL/well of the protease solution and shake the plates at 750 rpm in a ATR Multitron shaker incubator or similar overnight.

    c. The next morning, add the appropriate amount of neutralization buffer.

12. Wash an appropriate amount of M2 M-280 magnetic beads three times in TBST by magnetic separation. Resuspend in the original volume of TBST and dispense into an Assay Block 1 mL 96-well plate at 200 µL/well. Remove the TBST after magnetic separation. Resuspend the beads in the protease treated phage and incubate with vortex mixing for 3 h. Wash three times with 100 µL/well of 5 mM Tris, 0.1 mg/mL BSA by magnetic separation. Wash three times with 100 µL/well of 5 mM Tris by magnetic separation. Remove the residual liquid.

13. Elute the bound peptides by addition of 10 µL/well of 0.1% TFA and vortex mixing for 5 min at room temperature.

14. Prepare the matrix solution by mixing 10 mg DHB with 0.333 mL acetonitrile and 0.667 mL 0.1% TFA. Add 5 µL of the matrix solution to a v-bottom polypropylene plate.

15. Remove 5 µL of the eluted peptides after magnetic separation and mix with the matrix solution in the v-bottom polypropylene plate.

16. Place 1 µL of the peptide/matrix solution per spot of an AnchorChip 400 target. Let dry at room temperature.

17. Determine the masses of the released peptides by MALDI TOF.

18. Determine the scissile bond position by matching the mass of the released peptide to the amino acid sequence obtained by translation of the DNA sequencing data of the PCR products (**Fig. 4**).

3.5. Data Analysis

The simplest and at the same time the most informative way of representing positional preferences of a given amino acid in a binding subsite is to use graphical sequence logos. This method was developed by Scheider and Stephens *(14)* and was implemented as a Web service, called WebLogo, by Crooks et al. *(15)*. It is available at: http://weblogo.berkeley.edu/. It displays the nucleic acid bases or amino acids patterns as a result of multiple sequence alignment. The patterns consist of stacks of characters representing the sequence. Each stack represents a single position in the sequence. The characters that occur most commonly are located at the top of the stack. The height of each character is proportional to its frequency at that position. The overall height of the stack reflects the sequence conservation at that position. Instead of using frequencies as a way of scaling the height of the entire stack, that height can be adjusted according to the information content (measured in bits) of the sequence at that position.

According to Schneider and Stephens *(14)*, the information content is calculated in the following way. First, the sequence

ADVGGTDYKDDDDKPGGRPTWPSSGGSGRSLSR↓LTAS

Calculated monoisotopic mass = 3394

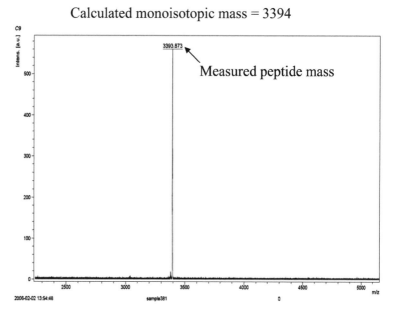

Fig. 4. Mapping of the position of the scissile bond by MALDI TOF MS. The amino acid sequence was determined by DNA sequencing of a PCR amplicons from an MMP-2 substrate phage infected bacterial culture followed by translation of the DNA sequence. The mass of the peptide released from 10^{12} MMP-2 substrate phage by MMP-2 was determined by MALDI TOF MS performed using Bruker Daltonics UltraFlex II mass spectrometer. (*See Color Plates*)

conservation at a specific position in the alignment: R_{seq} is calculated as the difference between the maximum possible entropy and the entropy of the observed symbol distribution:

$$R_{seq} = S_{max} - S_{obs} = \log_2 N - \left(-\sum_{n=1}^{N} p_n \log_2 p_n \right)$$

where p_n is the observed frequency of symbol n at a specific position, N is a number of distinct symbols used in a sequence type, which is 4 for nucleic acids and 20 for proteins., yielding the maximum value of entropy ($\log_2 N$) equal to 2 or 4.32, for nucleic acids and proteins, respectively. The sum of sequence conservation (R_{seq}) at each position of the logo measures the information content of the logo. This is true under the assumption that there is no inter-positional correlation and the background distribution of symbols is uniform.

In our calculation for determining patterns in the set of substrates for matrix metalloproteinases studied in the phage display experiments, we use the frequency-based instead of information-based logos. To prepare input for the WebLogo program, e.g., the set of aligned sequences, we use an approach that is based on Position Weight Matrix (PWM). A PWM assumes independence between positions in the pattern and allows one to take into account the background distribution. An element in a PWM is calculated as $m_{i,j} = \log(p_{i,j}/b_i)$, where $p_{i,j}$ is the probability of observing symbol (amino acid) i at position j of the motif, and b_i is the probability of observing the symbol i in a background model. The value of particular element of the PWM is the log-odds of a given symbol being found in the current pattern vs. being found in the background distribution. From the PWM matrix one can rederive the set of sequences that correspond to appropriate log-odds values, which in turn can be used as input to WebLogo program.

The specific procedure of deriving PWM for substrates of metalloproteinases is as follows. First, we aligned the sequences of all substrates for a given enzyme along the hydrolyzed peptide bond. Each sequence of the substrate in the phage display experiment consists of six amino acids variable regions and the N-terminal and C-terminal constant tags. Since a hydrolysis site can occur at any position in the six amino acids variable region, we included some portion of the constant tags in the sequence alignment. All aligned sequences have been uniformly trimmed to the length of eight amino acids; only those amino acids that span P5-P3 positions have been left for further consideration. In the next step, we derived the position weight matrices (PWMs), specific for each metalloproteinase, using the frequencies of each individual amino acid at each position in the alignment. The PWMs have been normalized by distribution of amino acids in the background sequences. In deriving distribution of the background sequences we included also short, three and two amino acid long fragments of the constant tags at the N- and C-terminals, respectively. In this way, we took into account biasing of the background and substrate distributions by constant fragments of the sequences. The background sequences are chosen randomly and are predicted to be hydrolyzed with less than 5% probability. In the last step, each position of the PWM is renormalized to the values in the range between 0 and 1 over all amino acids. Then each value for a given amino acid at a given position is multiplied by 100. This is the final number each amino acid would appear in the set of sequences, which are used as input for the WebLogo program. An example of the PWM for the 234 substrates of MMP-16 metalloproteinase and the corresponding sequence logo are shown in **Figs. 5** and **6**.

3.6. Concluding Remarks

Protease substrate phage display is a powerful tool for delineation of protease substrate specificity at the level of amino acid sequence. Our group adapted this approach for high throughput data acquisition and analysis, thereby making it capable of studying

	P5	P4	P3	P2	P1	P1'	P2'	P3'
G	1.0443	1.0953	0.6594	0.7880	1.2216	0.0000	0.0000	0.9380
A	1.1629	1.1686	2.5989	2.4216	3.0451	0.0000	0.1562	2.2099
P	1.7153	0.6433	6.2323	0.0000	1.7865	0.0000	0.0000	0.0000
C	0.0000	0.7773	0.0000	0.0000	0.5263	0.0000	0.0000	0.0000
T	0.7203	0.7897	0.2410	0.2045	0.0000	0.0000	2.3218	0.9794
S	1.2667	0.6583	0.7563	0.8359	1.8858	0.0000	0.4945	0.6868
D	0.4686	0.0000	0.0000	0.1799	0.1799	0.0000	0.0000	0.2349
N	1.3567	0.9567	0.0000	0.4922	2.8299	0.0000	0.0000	0.5255
E	0.9691	1.8290	0.0000	0.5800	0.0000	0.0000	0.2858	0.6043
Q	0.5922	1.7768	0.0000	1.4764	0.3667	0.3667	0.9638	0.5907
K	0.3676	1.5645	0.0000	0.3790	0.1263	0.0000	1.2163	0.8699
H	0.9910	1.1307	0.4276	0.8938	5.5413	0.0000	0.0000	0.9001
R	1.2300	1.5049	0.0831	2.2120	0.2011	0.1340	3.1541	0.5378
V	0.9008	1.1364	2.6148	0.1858	0.0000	0.5572	0.5783	0.5983
I	0.4180	0.9013	0.8124	0.5337	0.0000	2.2787	0.8477	0.3116
L	0.5670	1.3711	2.4054	1.3978	0.0667	8.2521	0.4679	0.8476
M	0.6396	0.5876	1.3449	1.6718	2.3406	6.6864	2.5044	1.3941
F	0.1904	0.6862	2.1288	1.8252	1.3037	0.5214	0.7187	0.5800
Y	1.4578	1.3267	0.7444	1.2860	1.4147	0.9001	0.8065	0.4230
W	0.2632	0.4564	1.5908	2.3544	1.6817	2.3541	1.4311	0.0000

Fig. 5. Position weight matrix of 234 substrates of MMP-16.

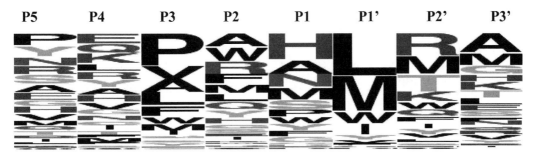

Fig. 6. WebLogo representation of the position weight matrix depicted in **Fig. 5**. (*See Color Plates*)

the evolution of substrate specificity in closely related families of proteases. These studies may shed light on the biological processes affected by different members of the family as a result of differential substrate specificities. As a bonus, one gets information on unique and selective peptide substrates that can be used for such applications as activity-based probes, selective inhibitors, and so on. We feel that the ambitious goal of characterizing substrate specificities of all human proteases is now well within reach. This work was supported by Grant Number RR020843 from the National Center for Research Resources (NCRR), a component of the National Institutes of Health (NIH) and its contents are solely the responsibility of the authors and do not necessarily represent the official view of NCRR or NIH.

References

1. Barrett AJ, Rawlings, Neil D., Woessner, J. Fred. (2004) *Handbook of Proteolytic Enzymes.* San Diego: Elsevier Academic Press.

2. Berger A, Schechter I. (1970) Mapping the active site of papain with the aid of peptide substrates and inhibitors. *Philos Trans R Soc Lond B Biol Sci,* 257(813), 249–64.

3. Diamond SL. (2007) Methods for mapping protease specificity. *Curr Opin Chem Biol* 11(1), 46–51.

4. Smith GP. (1985) Filamentous fusion phage: novel expression vectors that display cloned antigens on the virion surface. *Science* 228(4705), 1315–7.

5. Matthews DJ, Wells JA. (1993) Substrate phage: selection of protease substrates by monovalent phage display. *Science* 260(5111), 1113–7.

6. Ohkubo S, Miyadera K, Sugimoto Y, Matsuo K, Wierzba K, Yamada Y. (1999) Identification of substrate sequences for membrane type-1 matrix metalloproteinase using bacteriophage peptide display library. *Biochem Biophys Res Commun* 266(2), 308–13.

7. Cloutier SM, Chagas JR, Mach JP, Gygi CM, Leisinger HJ, Deperthes D. (2002) Substrate specificity of human kallikrein 2 (hK2) as determined by phage display technology. *Eur J Biochem* 269(11), 2747–54.

8. Smith MM, Shi L, Navre M. (1995) Rapid identification of highly active and selective substrates for stromelysin and matrilysin using bacteriophage peptide display libraries. *J Biol Chem* 270(12), 6440–9.

9. Hills R, Mazzarella R, Fok K, et al. (2007) Identification of an ADAMTS-4 cleavage motif using phage display leads to the development of fluorogenic peptide substrates and reveals matrilin-3 as a novel substrate. *J Biol Chem* 282(15), 11101–9.

10. Pan W, Arnone M, Kendall M, et al. (2003) Identification of peptide substrates for human MMP-11 (stromelysin-3) using phage display. *J Biol Chem* 278(30), 27820–7.

11. Deng SJ, Bickett DM, Mitchell JL, et al. (2000) Substrate specificity of human collagenase 3 assessed using a phage-displayed peptide library. *J Biol Chem* 275(40), 31422–7.

12. Kridel SJ, Chen E, Kotra LP, Howard EW, Mobashery S, Smith JW. (2001) Substrate hydrolysis by matrix metalloproteinase-9. *J Biol Chem* 276(23), 20572–8.

13. Bieth JG. (1995) Theoretical and practical aspects of proteinase inhibition kinetics. *Methods Enzymol* 248, 59–84.

14. Schneider TD, Stephens RM. (1990) Sequence logos: a new way to display consensus sequences. *Nucleic Acids Res* 18(20), 6097–100.

15. Crooks GE, Hon G, Chandonia JM, Brenner SE. (2004) WebLogo: a sequence logo generator. *Genome Res* 14(6), 1188–90.

Chapter 7

Imaging Specific Cell Surface Protease Activity in Living Cells Using Reengineered Bacterial Cytotoxins

John P. Hobson, Shihui Liu, Stephen H. Leppla, and Thomas H. Bugge

Summary

The scarcity of methods to visualize the activity of individual cell surface proteases in situ has hampered basic research and drug development efforts. In this chapter, we describe a simple, sensitive, and noninvasive assay that uses nontoxic reengineered bacterial cytotoxins with altered protease cleavage specificity to visualize specific cell surface proteolytic activity in single living cells. The assay takes advantage of the absolute requirement for site-specific endoproteolytic cleavage of cell surface-bound anthrax toxin protective antigen for its capacity to translocate an anthrax toxin lethal factor-β-lactamase fusion protein to the cytoplasm. A fluorogenic β-lactamase substrate is then used to visualize the cytoplasmically translocated anthrax toxin lethal factor-β-lactamase fusion protein. By using anthrax toxin protective antigen variants that are reengineered to be cleaved by furin, urokinase plasminogen activator, or metalloproteinases, the cell surface activities of each of these proteases can be specifically and quantitatively determined with single cell resolution. The imaging assay is excellently suited for fluorescence microscope, fluorescence plate reader, and flow cytometry formats, and it can be used for a variety of purposes.

Key words: Anthrax toxin, β-lactamase, CCF2/AM, Cell surface proteolysis, Flow cytometry, Fluorescence microscopy, Fluorescence plate reader, Furin, Metalloproteinases, Urokinase plasminogen activator.

1. Introduction

Proteolysis in the extracellular environment is essential to all aspects of life, including development, homeostasis, tissue repair, tissue remodeling, and reproduction. Dysregulated extracellular proteolysis is also causally linked to the genesis or progression of a large number of human diseases. Important examples include myocardial infarction, stroke, rheumatoid and osteoarthritis,

Toni M. Antalis and Thomas H. Bugge (eds.), *Methods in Molecular Biology, Proteases and Cancer, Vol. 539*
© Humana Press, a part of Springer Science+Business Media, LLC 2009
DOI: 10.1007/978-1-60327-003-8_7

periodontal disease, bacterial infection, and cancer. Physiological and pathological proteolysis within the extracellular environment takes place predominantly on the cell surface, where inactive protease zymogens, protease receptors, protease inhibitors, and other regulatory proteins assemble into complex multiprotein structures that mediate the sequential and spatially restricted activation of protease zymogens and their subsequent cleavage of target proteins (1–7).

Despite its critical physiological and pathological importance, several fundamental aspects of extracellular proteolysis are still quite poorly understood. These include the molecular pathways that initiate the activation and inhibition of many zymogen cascades and the biologically relevant substrates for many, if not the majority, of pericellular proteases. In particular, the lack of imaging agents that can specifically visualize the endogenous activity of individual proteases and monitor the efficacy of their pharmacological inhibition in situ is widely recognized as a serious impediment to both basic research efforts and to the use of proteases as therapeutic targets (8, 9). Considerable efforts therefore have been expended to develop assays for the imaging of specific protease activity in biological systems, and a wide variety of approaches have been taken (10, 11).

In this chapter, we describe a simple noninvasive method for imaging specific cell surface proteolytic activity in single living cells that is based on the use of modified bacterial cytotoxins (12). The assay takes advantage of: (a) the absolute requirement for site-specific endoproteolytic cleavage of anthrax toxin protective antigen (PrAg) for toxin activation (13), (b) the ability to reengineer the protease cleavage site in PrAg to be specifically cleaved by various different cell surface-associated proteases (14–18), (c) the ability of anthrax toxin lethal factor (LF) to shuttle heterologous proteins to the cytoplasm in a PrAg cleavage-dependent manner (19, 20), and (d) the development of highly sensitive optical assays for the detection of cytoplasmic β-lactamase activity (21). The imaging assay combines high protease specificity – obtained by the incorporation of the target peptide cleavage sequence in a macromolecular context and the confinement of the cleavage reaction to the cell surface (16, 22) – with the exquisite sensitivity that is associated with enzymatic signal amplification via β-lactamase (21). We provide protocols for the application of the assay for fluorescence microscopy, fluorescent plate reader, and flow cytometry formats.

The devised assay can be tailored to detect the activity or inhibition of any cell surface protease for which a specific peptide substrate can be derived (16, 22–25). Examples of applications for which the imaging assay is particularly suited include: (1) automated high-throughput screening of large chemical libraries for protease inhibitory or stimulating compounds (26); (2) expression cloning-based

Color Plates

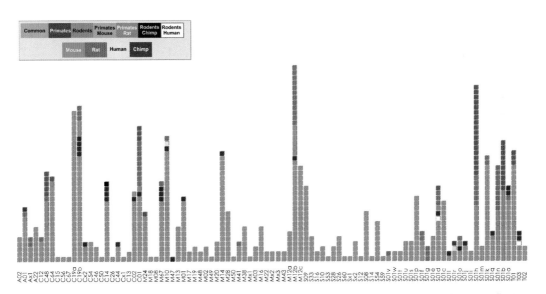

Chapter 2, Fig. 2. Comparative analysis of human, chimpanzee, mouse, and rat degradomes. The complete nonredundant set of proteases and protease homologues from each species is distributed in five catalytic classes and 68 families. Each *square* represents a single protease and is *colored* according to its presence or absence in human, chimpanzee, mouse, and rat as indicated in the *inset*.

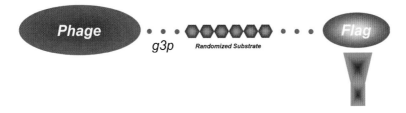

Chapter 6, Fig. 1. The affinity tagged randomized peptide substrate fusion protein with the gene 3 protein (g3p) of filamentous M13 bacteriophage.

ADVGGTDYKDDDDKPGGRPTWPSSGGSGRSLSR↓LTAS

Calculated monoisotopic mass = 3394

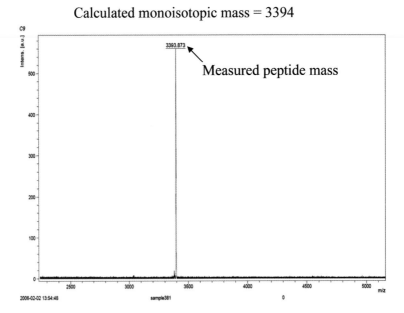

Chapter 6, Fig. 4. Mapping of the position of the scissile bond by MALDI TOF MS. The amino acid sequence was determined by DNA sequencing of a PCR amplicons from an MMP-2 substrate phage infected bacterial culture followed by translation of the DNA sequence. The mass of the peptide released from 10^{12} MMP-2 substrate phage by MMP-2 was determined by MALDI TOF MS performed using Bruker Daltonics UltraFlex II mass spectrometer.

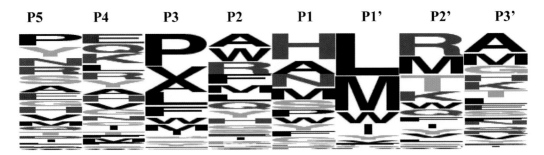

Chapter 6, Fig. 6. WebLogo representation of the position weight matrix depicted in **Fig. 5**.

strategies for the identification of new activators and inhibitors of specific cell surface proteolytic pathways; (3) siRNA and shRNA-based high-throughput screens for the identification of pathways regulating specific cell surface protease activity.

2. Materials

2.1. Modified Anthrax Toxins

1. The generation and purification of recombinant wild-type PrAg, PrAg-33 ([164]RKKR[167] of PrAg changed to RAAR) *(14)*, PrAg-U2 ([164]RKKR[167] of PrAg changed to GSGRSA) *(17, 22)*, PrAg-L1 ([164]RKKR[167] of PrAg changed to GPLGMLSQ), and PrAg-U7 ([164]RKKR[167] changed to GG) *(16)* is described in Chapter 10, Dissecting the Urokinase Activation Pathway Using Urokinase-Activated Anthrax Toxin, page (3–4 of Chapter). Store at –80°C in aliquots of 50–500 µL at concentrations >500 µg/mL.

2. The LF/β-lactamase fusion protein (LF/β-Lac) contains DNA sequences encoding the PrAg-binding region of LF (amino acids 1–254) fused 5′ of DNA sequences encoding amino acids 19–286 of the TEM-1 β-lactamase gene from pBluescript (Stratagene, La Jolla, CA). To generate LF/β-Lac, these DNA sequences were fused downstream of a DNA sequence encoding a glutathione-*S*-transferase (GST) protein with a thrombin cleavage site linking GST and LF/β-Lac *(27, 12)*. BL21 competent bacteria (Invitrogen, Carlsbad, CA) transformed with the expression plasmid encoding the GST- LF/β-Lac fusion protein are grown in Super Broth Medium (Biosource, Camarillo, CA) supplemented with 50 µg/mL ampicillin. Cultures are grown at 37°C until an absorbance of 0.8 at 600 nm. The cells are then induced with isopropyl β-D-thiogalactopyranoside (Sigma, St. Louis, MO) to a final concentration of 0.5 mM for 5 h at 37°C. The cells are pelleted by centrifugation at 3,000 × g for 15 min in a 4°C high speed centrifuge and resuspended in ice-cold bacterial lysis buffer containing 50 mM Tris-HCl, pH 7.5, 150 mM NaCl, 1% Triton X-100, and a protease inhibitor cocktail (Amersham Biosciences, Piscataway, NJ). Bacterial pellets are lysed by five rounds of freeze/thawing in liquid nitrogen, followed by sonication for 30 s using a probe sonicator. The lysate is clarified by centrifugation 3,000 × g for 10 min in a 4°C high-speed centrifuge and incubated with 1 mL glutathione-Sepharose resin slurry (Amersham Biosciences) for 2 h at 4°C. The resin is washed twice with ice-cold lysis buffer without protease inhibitors, and the purified protein is released from the resin by incubation

with 2 mL thrombin cleavage buffer (20 mM Tris-HCl, pH 8.4, 150 mM NaCl, 2.5 mM CaCl$_2$ containing 100 Units thrombin/mL). Aliquots of 0.5 mg/mL LF/β-Lac are stored at –80°C.

2.2. Tissue Culture and Imaging Assay

1. Phenol red-free DMEM (Gibco-BRL, Gaithersburg, MD) is stored at 4°C.

2. Gentamicin reagent solution is purchased as a liquid 50× solution from Gibco-BRL. Store at 4°C.

3. L-Glutamine is purchased as a 100× solution from Gibco-BRL. Store at –20°C.

4. 0.05% trypsin 1 mM EDTA solution is purchased as a 1× solution from Gibco-BRL and stored at 4°C.

5. Poly-D-lysine is purchased from Sigma. Make a sterile 0.1 mg/mL solution in PBS. Store at 4°C.

6. Ten cm tissue culture plates, eight-well chamber slides, and 96-well black walled tissue culture plates are purchased from Corning, Lowell, MA.

7. Hanks' balanced salt solution is purchased as a sterile 1× solution from Sigma. Store at 4°C.

8. Probenecid is purchased from Sigma.

9. Human pro-uPA (single-chain uPA, no. 107) and human Glu-plasminogen is purchased from American Diagnostica, Inc. (Greenwich, CT).

2.3. β-Lactamase Substrate and Substrate Loading Solutions

1. "Solution A" [Coumarin cephalosporin fluorescein acetoxymethyl ester (CCF2/AM) in dimethyl sulfoxide (DMSO)]. CCF2/AM is purchased as a dried powder from Invitrogen (Carlsbad, CA). Store at –80°C and protect from light. Dissolve 1 mg CCF2/AM in 924 μL DMSO. Store protected from light in 100 μL aliquots at –80°C. Bring the vial of CCF2/AM to room temperature before removing the desired amount of reagent to reduce intrusion of moisture into the stock solution.

2. "Solution B" (100 mg/mL Pluronic®-F127 surfactant in DMSO with 0.1% acetic acid) is purchased from Invitrogen. Store at room temperature. Protect from direct light.

3. "Solution C" (24% w/w PEG 400, 18% TR-40 w/w in water) is purchased from Invitrogen. Store at room temperature. Protect from direct light.

4. Preparation of 6× "Standard substrate loading solution" (must be prepared immediately before use): In an opaque test tube, add 1 volume "Solution A" to 5 volumes "Solution B" and vortex. Add the combined "Solution A" and

"Solution B" to 77 volumes "Solution C" and vortex. Reagents should be exposed to as little direct light as possible.

5. Preparation of 6× "Alternative substrate loading solution": Dissolve 25 g probenecid in 219 mL of 400 mM NaOH in a flask using a stir bar. Add 219 mL of sodium phosphate buffer (pH 8.0). Adjust the pH to 8.0 using 1 M HCl and/ or 1 M NaOH. If a precipitate forms, bring into solution by continuous stirring. Freeze the probenecid solution in 1 mL aliquots at −20°C. Immediately before use in an opaque test tube, add 1 volume "Solution A" to 5 volumes "Solution B" and vortex. Add the combined "Solution A" and "Solution B" to 77 volumes "Solution C" and vortex. Add 4 volumes probenecid solution to combined "Solution A," "Solution B," and "Solution C." Reagents should be exposed to as little direct light as possible.

3. Methods

The principle of the imaging assay is shown schematically in **Fig. 1**. Live cell imaging of furin (**Fig 2**), uPA (**Fig. 3**), and metalloproteinase activity (**Fig 4**) can be performed using a wide variety of cultured primary cells and cell lines, by treating the cells with PrAg-33, PrAg-U2, or PrAg-L1, respectively, in combination with LF/β–Lac. Cleavage of PrAg-33, PrAg-U2, or PrAg-L1 by, respectively, furin, uPA, or metalloproteinases allows the translocation of LF/β-Lac to the cytoplasm. Cytoplasmic LF/β–Lac is imaged by the incubation of cells with the fluorogenic β-lactamase substrate, CCF2/AM, and detection of hydrolyzed CCF2/AM by a fluorescence microscope, a fluorescence plate reader, or a flow cytometer. CCF2/AM fluoresces green before its hydrolysis by β-lactamase and fluoresces blue after hydrolysis by β-lactamase. The noncleavable PrAg variant, PrAg-U7 combined with LF/β–Lac serves as a negative control, and wild-type PrAg, which is cleaved by furin as well as many furin-like proprotein convertases, combined with LF/β–Lac serves as a positive control. The basic protocols for imaging cell surface protease activity by fluorescence microscopy, fluorescence plate reader, and flow cytometry are given below. The cells can be subjected to a large variety of manipulations and treatments prior to imaging and during the imaging process depending on the specific purpose of the experiment (*see* **Subheading 1**).

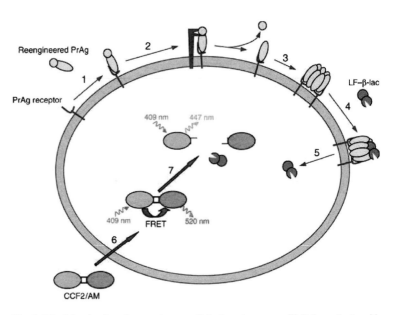

Fig. 1. Principle of cell surface protease activity imaging assay. (1) PrAg mutants with reengineered protease cleavage specificity bind to ubiquitous high affinity cell surface receptors. (2) The PrAg mutants are cleaved by the cell surface proteolytic activity to be imaged. (3) The PrAg fragment that remains on the cell surface heptamerizes. (4) High-affinity binding sites for LF/β-Lac are generated by heptamerization of the PrAg. (5) LF/β-Lac is translocated to the cytoplasm after endocytosis of the PrAg-LF/β-Lac complex (not shown). (6) CCF2/AM is added to cells and diffuses into the cytoplasm where it is trapped by hydrolysis of its hydroxethyl ester groups by nonspecific cytoplasmic esterases. (7) LF/β-Lac hydrolyses the cephalosporin ring of CCF2/AM, causing a shift in fluorescence emission from 520 nm (green light) to 447 nm (blue light) after excitation of cells at 409 nm. Blue fluorescence emission by a cell demonstrates specific cell surface proteolytic activity. Reproduced from **ref.** *12.*

3.1. Imaging Cell Surface Proteolytic Activity by Fluorescence Microscopy Analysis

1. Grow cells to be imaged in phenol red-free DMEM supplemented with 10% fetal bovine serum (FBS), L-glutamine, and gentamicin reagent solution (*see* **Note 1**). Avoid using penicillin and other β-lactam-based antibiotics that are substrates for β-lactamase.

2. Treat eight-well chamber slides with 250 μL/well 0.1 mg/mL poly-D-lysine in PBS at 37°C in a tissue culture incubator for 60 min to facilitate cell adhesion. Other treatments of chamber slides can be used as is expedient for the adhesion of the cell line or primary cells employed (*see* **Note 2**).

3. Seed cells on the eight-well chamber slides the day prior to imaging in phenol red-free DMEM supplemented with 10% FBS, L-glutamine, and gentamicin reagent solution. The seeding of 2×10^5 cells/well is appropriate for most cells

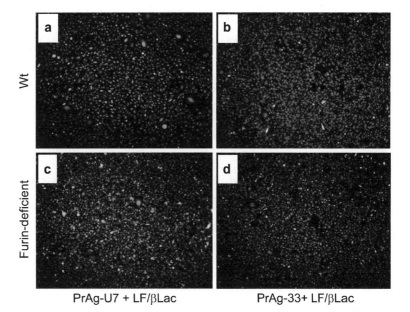

Fig. 2. Specific imaging of endogenous cell surface furin proteolytic activity in single living cells. Wild-type (**a** and **b**) or furin-deficient (**c** and **d**) Chinese hamster ovary cells were incubated with 90 nM LF/β-Lac and 26 nM (2 μg/mL) PrAg-U7 (**a** and **c**), or 90 nM LF/β-Lac and 26 nM PrAg-33 (**b** and **d**) for 60 min. Thereafter, 1.5 μM CCF2/AM was added to the cells for 60 min at room temperature. The CCF2/AM remaining in the medium was removed by washing, and the cells were incubated for 60 min at room temperature to allow for CCF2/AM hydrolysis. The cells then were subjected to fluorescence microscopy using an excitation wavelength of 405 nm and emission filters of 530 nm (green light) and 460 nm (blue light). Specific imaging of furin proteolytic activity is demonstrated by the bright blue fluorescence of wild-type cells (**b**), but not furin-deficient cells (**d**), treated with LF/β-Lac and PrAg-33, or wild-type and furin-deficient cells treated with LF/β-Lac and PrAg-U7 (**a** and **c**). Reproduced in part from **ref.** *12*.

(*see* **Note 3**). Incubate overnight in a standard 37°C tissue culture incubator.

4. Gently wash the cells twice in 0.5 mL 37°C phenol red-free DMEM.

5. Add either PrAg-33, PrAg-U2, or PrAg-L1 to 26 nM (2 μg/mL) final concentration together with LF/β-Lac fusion protein to 90 nM (5 μg/mL) final concentration in 0.25 mL 37°C phenol red-free DMEM. When imaging uPA cell surface proteolytic activity, include 1 μg/mL Glu-plasminogen with or without 100 ng/mL pro-uPA (*see* **Note 4**). As a negative control, substitute PrAg-33, PrAg-U2, or PrAg-L1 with PrAg-U7. As a positive control, substitute PrAg-33, PrAg-U2, or PrAg-L1 with wild-type PrAg (*see* **Note 5**).

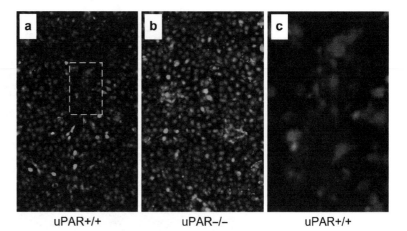

uPAR+/+   uPAR–/–   uPAR+/+

Fig. 3. Specific imaging of endogenous cell surface uPA activity in mixed tumor cell stromal cell cultures. Mixed cultures of HN-26 oral squamous carcinoma cells and wild type (uPAR+/+, **a** and **c**) or urokinase plasminogen activator receptor (uPAR) knockout (uPAR$^{-/-}$, **b**) fibroblasts were treated with 90 nM LF/β-Lac (5 μg/mL), 26 nM (2 μg/mL) PrAg-U2, and 11 nM (1 μg/mL) plasminogen for 60 min. Thereafter, 1.5 μM CCF2/AM was added to the cells for 60 min at room temperature. The CCF2/AM remaining in the medium was removed by washing, and the cells were incubated for 60 min at room temperature to allow for CCF2/AM hydrolysis. The cells then were subjected to fluorescence microscopy using an excitation wavelength of 405 nm and emission filters of 530 (green light) and 460 nm (blue light). Cell surface uPA activity is exclusively observed in nontransformed fibroblasts (**a** and **c**) and is dependent on the expression of uPAR, as shown by the strong blue fluorescence of islands of uPAR+/+ (**a**), but not in uPAR–/– (**b**) fibroblasts, and by green fluorescence of the tumor cells. (**c**) is a high magnification of the boxed area in (**a**) illustrating the confinement of uPA activity to fibroblasts. Reproduced in part from **ref.** *12*.

6. Incubate the PrAg and LF/β-Lac-treated cells for 60 min at 37°C in a tissue culture incubator.

7. Wash cells twice with room temperature phenol red-free DMEM. Transfer plate to room temperature (*see* **Note 6**).

8. Add 40 μL of 6× "Standard substrate loading solution" or "Alternative substrate loading solution" (*see* **Subheading 2.3**) to each chamber well containing 210 μL of room temperature phenol red-free DMEM. The use of standard or alternative loading solutions is cell type-dependent (*see* **Note 7**).

9. Cover the plate with aluminum foil and incubate for 60 min at room temperature to allow the loading of cells with CCF2/AM (*see* **Note 8**).

10. Wash the cells three times with room temperature phenol red-free DMEM and incubate for additional 60 min at room temperature to allow hydrolysis of CCF2/AM by the cytoplasmically-translocated LF/β-Lac (*see* **Note 9**).

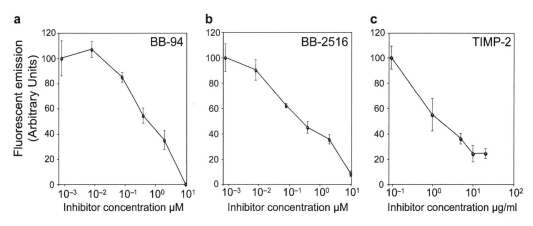

Fig. 4. Quantitative analysis of the inhibition of cell surface metalloproteinase activity in human tumor cells. HT-1080 fibrosarcoma cells were seeded in a 96-well microtiter plate and treated for 60 min with 90 nM (5 μg/mL) LF/β-Lac and 26 nM (2 μg/mL) PrAg-L1 in the presence of increasing concentrations of the metalloprotease inhibitors BB-94 (**a**), BB-2516 (**b**), and TIMP-2 (**c**). Thereafter, 1.5 μM CCF2/AM was added to the cells for 60 min at room temperature. The CCF2/AM remaining in the medium was removed by washing, and the cells were incubated for 60 min at room temperature to allow for CCF2/AM hydrolysis. Fluorescence emission was recorded with a plate reader using 405 nm excitation and 460/25 nm bandpass for blue fluorescence and 535/25 nm bandpass for green fluorescence. The data are expressed as mean ± standard error of the mean of triplicate determinations. Reproduced in part from **ref.** *12*.

11. Inspect cells using an inverted fluorescence microscope such as a Zeiss Axioplan (Carl Zeiss, Jena, Germany). For acquisition of blue fluorescence with this microscope, use excitation filter HQ405/20 nm bandpass, dichroic 425DCXR, and emitter filter HQ460/40 nm bandpass. For acquisition of green fluorescence with this microscope, use HQ405/20 nm bandpass, dichroic 425DCXR, and emitter filter HQ530/30 nm bandpass from Chroma Technology (Rockingham, VT) (*see* **Note 10**).

3.2. Imaging Cell Surface Proteolytic Activity with a Fluorescence Plate Reader

1. Grow cells to be imaged in phenol red-free DMEM (Gibco-BRL, Gaithersburg, MD) supplemented with 10% FBS, glutamine, and gentamicin reagent solution (*see* **Note 1**). Avoid using penicillin and other β-lactam-based antibiotics that are substrates for β-lactamase.

2. Treat 96-well black wall plates with a clear bottom with 150 μL/well 0.5 mg/mL poly-D-lysine in PBS at room temperature for 60 min to facilitate cell adhesion (*see* **Note 2**).

3. Seed cells 24 h before the assay at a density of 2×10^4 cells/well in 150 μL phenol red-free DMEM supplemented with 10% fetal bovine serum (FBS), glutamine, and gentamicin reagent solution (*see* **Note 3**). Add 150 μL medium without cells to six wells (*see* **step 14**).

4. Incubate plates over-night in a standard 37°C tissue culture incubator.

5. Gently wash the wells twice with 150 μL 37°C phenol red-free DMEM.

6. Add 26 nM (2 μg/mL) PrAg variant and 90 nM (5 μg/mL LF/β-Lac) fusion protein in 120 μL 37°C phenol red-free DMEM to each well. When imaging cell surface uPA proteolytic activity include 1 μg/mL Glu-plasminogen with or without 100 ng/mL pro-uPA (*see* **Note 4**). As a negative control, substitute PrAg-33, PrAg-U2, or PrAg-L1 with PrAg-U7. As a positive control, substitute PrAg-33, PrAg-U2, or PrAg-L1 with wild type PrAg (*see* **Note 5**).

7. Incubate the PrAg and LF/β-Lac-treated plate for 60 min at 37°C in a tissue culture incubator.

8. Wash cells twice with 150 μL room temperature phenol red-free DMEM. Transfer the plate to room temperature (*see* **Note 6**).

9. Add 20 μL of 6× "Standard substrate loading solution" or "Alternative substrate loading solution" (*see* **Subheading 2.3**) in 100 μL room temperature phenol red-free DMEM to each well. The use of standard or alternative loading solutions is cell type-dependent (*see* **Note 7**).

10. Cover the plate with aluminum foil to protect CCF2/AM from the light and incubate for 60 min at room temperature to allow the loading of cells with CCF2/AM (*see* **Note 8**).

11. Wash the cells three times with 150 μL room temperature phenol red-free DMEM containing Probenicid and incubate for an additional 60 min at room temperature to allow hydrolysis of CCF2/AM by the cytoplasmically-translocated LF/β-Lac (*see* **Note 9**).

12. Remove dust from the bottom of the plate.

13. Collect data using a fluorescence plate reader, such as a VICTOR plate reader from Perkin Elmer, Wellesley, MA, set in the bottom-read mode. Use the following filters in dual read mode: excitation 405/10 nm bandpass, emission filters 460/25 nm bandpass for blue fluorescence, and 535/25 nm bandpass for green fluorescence (*see* **Note 10**).

14. Determine "average blue background" and "average green background" from the wells without cells.

15. Subtract the "average blue background" and the "average green background" from all wells with cells to obtain "net blue signal" and "net green signal" for each well.

16. Divide "net blue signal" with "net green signal" for each well to obtain the "response ratio" (*see* **Note 11**).

3.3. Imaging Cell Surface Proteolytic Activity in Individual Cells by Flow Cytometry

1. Grow cells to be imaged in phenol red-free DMEM supplemented with 10% FBS, glutamine, and gentamicin reagent solution (*see* **Note 1**). Avoid using penicillin and other β-lactam ring-based antibiotics that are substrates for β-lactamase.

2. Treat 10 cm standard tissue culture dishes with 2 mL 0.1 mg/mL poly-D-lysine in PBS at 37°C in a tissue culture incubator for 60 min to facilitate cell adhesion (*see* **Note 2**).

3. Seed cells in the 10 cm tissue culture dishes 24 h before at a density of 1×10^6 cells/plate in phenol red-free DMEM supplemented with 10% FBS, glutamine, and gentamicin reagent solution (*see* **Note 3**).

4. Incubate overnight in a standard 37°C tissue culture incubator.

5. Gently wash the plates twice with 3 mL 37°C phenol red-free DMEM.

6. Add 26 nM (2 μg/mL) PrAg variant and 90 nM (5 μg/mL LF/β-Lac) fusion protein in 4 mL 37°C phenol red-free DMEM. When imaging cell surface uPA proteolytic activity include 1 μg/mL Glu-plasminogen with or without 100 ng/mL pro-uPA (*see* **Note 4**). As a negative control, substitute PrAg-33, PrAg-U2, or PrAg-L1 with PrAg-U7. As a positive control, substitute PrAg-33, PrAg-U2, or PrAg-L1 with wild-type PrAg (*see* **Note 5**).

7. Incubate the PrAg and LF/β-Lac-treated plates for 60 min at 37°C in a tissue culture incubator.

8. Wash cells twice with 3 mL room temperature phenol red-free DMEM. Transfer plate to room temperature (*see* **Note 6**).

9. Add 4 mL of 6× "Standard substrate loading solution" or 6× "Alternative substrate loading solution" (*see* **Subheading 2.3**) in 4 mL room temperature phenol red-free DMEM to each well. The use of standard or alternative loading solutions is cell type-dependent (*see* **Note 7**).

10. Cover the plates with aluminum foil to protect CCF2/AM from the light and incubate for 60 min at room temperature to allow the loading of cells with CCF2/AM (*see* **Note 8**).

11. Wash the plates three times with 3 mL room temperature phenol red-free DMEM and incubate for additional 60 min at room temperature to allow hydrolysis of CCF2/AM by the cytoplasmically-translocated LF/β-Lac (*see* **Note 9**).

12. Remove the cells from the plate using 1 mL trypsin-EDTA solution. Neutralize trypsin with 1 mL phenol red-free DMEM with 10% FBS. Add to a 10 mL tube with 8 mL

ice-cold Hanks' balanced salt solution containing 2 mM probenecid (*see* **Note 12**).

13. Centrifuge the cells for 2 min at 700 × g at 4°C. Discard medium.

14. Gently resuspend cells in 10 mL ice-cold Hanks' balanced salt solution containing 2 mM probenicid.

15. Centrifuge the cells for 2 min at 700 × g at 4°C. Discard medium. Gently resuspend cells in ice-cold Hanks' balanced salt solution containing 2 mM probenecid at a concentration of 1×10^6 cells/mL.

16. Analyze cells using a flow cytometer such as a BD LSRII flow cytometer from Beckton Dickinson, San Jose, CA, equipped with a standard 488 nm laser and a 405 nm violet laser. Exclude dead cells, cell debris, and aggregates using side scatter versus forward scatter.

17. Use a 405 nm excitation filter, 525/50 nm filter for green light emission, and 440/40 nm filter for blue light emission with a 505 LP dichroic mirror to determine blue and green fluorescence signals from each living cell.

18. Determine the relative cell surface proteolytic activity of each cell by dividing the blue signal with the green signal for each individual cell (*see* **Note 13**). Plot cell number versus blue/green signal as a histogram.

4. Notes

1. Other phenol red-free media compositions, sera, and antibiotic combinations can be used for cultivation as required for each individual cell line/primary cell type.

2. Cell types that adhere well to untreated tissue culture plastic or glass during normal culturing conditions may dislodge from an uncoated surface during the imaging procedure. The need for coating should be determined empirically.

3. Loading efficiency of cells with CCF2/AM depends on cell density and has been reported by Invitrogen to be most efficient at 60–80% confluence. However, we have successfully imaged cell surface protease activity in highly confluent cultures.

4. Addition of plasminogen is essential to convert pro-uPA to active two-chain uPA if the assay is performed in serum-free medium. Add pro-uPA when imaging the capacity of cells to

convert exogenous pro-uPA to active uPA on the cell surface. Do not add pro-uPA when imaging endogenous uPA activity. Note that pro-uPA and uPA binding to the urokinase plasminogen activator receptor (uPAR) is species-specific. For example, human uPA will not bind to mouse uPAR and mouse uPA will not bind to human uPAR.

5. PrAg-U7 is a noncleavable PrAg variant that binds to cells but cannot support the translocation of LF/β-Lac to the cytoplasm. The combination of PrAg-U7 and LF/β-Lac is the optimal negative control. Wild-type PrAg is cleaved by furin as well as a number of furin-like proprotein convertases. The combination of wild-type PrAg and LF/β-Lac is the optimal positive control.

6. The uptake of CCF2/AM into cells is faster at 37°C than at room temperature, but the outwards transport of CCF2/AM is also higher. The net result of incubation of cells at 37°C is decreased loading of cells with CCF2/AM.

7. In our experience, the alternative substrate loading solution method from Invitrogen works best with cell lines, and the standard loading solution works best with primary cell cultures.

8. Loading with CCF2/AM for 60 min suffices for most cells. The loading time can be extended for higher sensitivity.

9. The substrate hydrolysis time can be extended to increase sensitivity.

10. Light of 409 nm wavelength causes maximum excitation of the coumarin group in CCF2/AM, leading to the emission of green light by the fluorescein group with maximum intensity at 520 nm due to fluorescent resonance energy transfer before hydrolysis by β-lactamase, and blue light with maximum intensity at 450 nm after hydrolysis by β-lactamase. The excitation and emission filters used here are those recommended by Invitrogen for the imaging of CCF2/AM. Other filters with similar properties can be used.

11. A ratiometric readout compensates for well-to-well variation in terms of cell number and CCF2/AM uptake and provides the most consistent data.

12. Probenecid is an anion transport inhibitor that reduces the outwards transport of CCF2/AM, CCF2, and probably hydrolyzed CCF2 from cells and increases the sensitivity of the assay.

13. A ratiometric readout compensates for variations in terms of cell size and CCF2/AM uptake and provides the most accurate data.

Acknowledgments

We thank Drs. Silvio Gutkind and Mary Jo Danton for critically reviewing this manuscript, and Drs. Kevin L. Holmes and David Stephany for expert assistance with flow cytometry. This work was supported by the NIH Intramural Research Program, by NIAID Support of Intramural Biodefense Research from ICs other than NIAID and by the Department of Defense (DAMD-17-02-1-0693) to Dr. Thomas H. Bugge. For imaging reagents, contact S. H. Leppla (*Leppla@nih.gov*) or T. H. Bugge (thomas.bugge@nih.gov).

References

1. Werb Z. (1997) ECM and cell surface proteolysis: regulating cellular ecology. *Cell* 91(4), 439–42.

2. Andreasen PA, Egelund R, Petersen HH. (2000) The plasminogen activation system in tumor growth, invasion, and metastasis. *Cell Mol Life Sci* 57(1), 25–40.

3. McCawley LJ, Matrisian LM. (2000) Matrix metalloproteinases: multifunctional contributors to tumor progression. *Mol Med Today* 6(4), 149–56.

4. Turk B, Turk D, Turk V. (2000) Lysosomal cysteine proteases: more than scavengers. *Biochim Biophys Acta* 1477(1–2), 98–111.

5. Koblinski JE, Ahram M, Sloane BF. (2000) Unraveling the role of proteases in cancer. *Clin Chim Acta* 291(2), 113–35.

6. Davie EW, Fujikawa K, Kisiel W. (1991) The coagulation cascade: initiation, maintenance, and regulation. *Biochemistry* 30(43), 10363–70.

7. Bugge TH. Proteolysis in carcinogenesis. (2003) In: Ensley JF, Gutkind JS, Jacob JR, Lippman SM Head and Neck Cancer. San Diego: Academic Press, 137–49.

8. Mohamed MM, Sloane BF. (2006) Cysteine cathepsins: multifunctional enzymes in cancer. *Nat Rev Cancer* 6(10), 764–75.

9. Coussens LM, Fingleton B. (2002) Matrisian LM. Matrix metalloproteinase inhibitors and cancer: trials and tribulations. *Science* 295(5564), 2387–92.

10. McIntyre JO, Matrisian LM. (2003) Molecular imaging of proteolytic activity in cancer. *J Cell Biochem* 290(6), 1087–97.

11. Sloane BF, Sameni M, Podgorski I, Cavallo-Medved D, Moin K. (2006) Functional imaging of tumor proteolysis. *Annu Rev Pharmacol Toxicol* 46, 301–15.

12. Hobson JP, Liu S, Rono B, Leppla SH, Bugge TH. (2006) Imaging specific cell-surface proteolytic activity in single living cells. *Nat Meth* 3(4), 259–61.

13. Duesbery NS, Vande Woude GF. (1999) Anthrax toxins. *Cell Mol Life Sci* 55(12), 1599–609.

14. Gordon VM, Klimpel KR, Arora N, Henderson MA, Leppla SH. (1995) Proteolytic activation of bacterial toxins by eukaryotic cells is performed by furin and by additional cellular proteases. *Infect Immun* 63(1), 82–7.

15. Adachi M, Kitamura K, Miyoshi T, et al. (2001) Activation of epithelial sodium channels by prostasin in Xenopus oocytes. *J Am Soc Nephrol* 12(6), 1114–21.

16. Liu S, Netzel-Arnett S, Birkedal-Hansen H, Leppla SH. (2000) Tumor cell-selective cytotoxicity of matrix metalloproteinase-activated anthrax toxin. *Cancer Res* 60(21), 6061–7.

17. Liu S, Aaronson H, Mitola DJ, Leppla SH, Bugge TH. (2003) Potent antitumor activity of a urokinase-activated engineered anthrax toxin. *Proc Natl Acad Sci USA* 100(2), 657–62.

18. Liu S, Schubert RL, Bugge TH, Leppla SH. (2003) Anthrax toxin: structures, functions and tumour targeting. *Expert Opin Biol Ther* 3(5), 843–53.

19. Leppla SH, Arora N, Varughese M. (1999) Anthrax toxin fusion proteins for intracellular delivery of macromolecules. *J Appl Microbiol* 187(2), 284.

20. Arora N, Leppla SH. (1993) Residues 1–254 of anthrax toxin lethal factor are sufficient to

cause cellular uptake of fused polypeptides. *J Biol Chem* 268(5), 3334–41.

21. Zlokarnik G, Negulescu PA, Knapp TE, et al. (1998) Quantitation of transcription and clonal selection of single living cells with beta-lactamase as reporter. *Science* 279(5347), 84–8.

22. Liu S, Bugge TH, Leppla SH. (2001) Targeting of tumor cells by cell surface urokinase plasminogen activator-dependent anthrax toxin. *J Biol Chem* 276(21), 17976–84.

23. Ke SH, Coombs GS, Tachias K, Navre M, Corey DR, Madison EL. (1997) Distinguishing the specificities of closely related proteases. Role of P3 in substrate and inhibitor discrimination between tissue-type plasminogen activator and urokinase. *J Biol Chem* 272(26), 16603–9.

24. Ke SH, Madison EL. (1997) Rapid and efficient site-directed mutagenesis by single-tube 'megaprimer' PCR method. *Nucleic Acids Res* 25(16), 3371–2.

25. Coombs GS, Bergstrom RC, Pellequer JL, et al. (1998) Substrate specificity of prostate-specific antigen (PSA). *Chem Biol* 5(9), 475–88.

26. Inglese J, Johnson RL, Simeonov A, et al. (2007) High-throughput screening assays for the identification of chemical probes. *Nat Chem Biol* 3(8), 466–79.

27. Arora N, Leppla SH. (1994) Fusions of anthrax toxin lethal factor with shiga toxin and diphtheria toxin enzymatic domains are toxic to mammalian cells. *Infect Immun* 62(11), 4955–61.

Chapter 8

Cell-Based Identification of Natural Substrates and Cleavage Sites for Extracellular Proteases by SILAC Proteomics

Magda Gioia, Leonard J. Foster, and Christopher M. Overall

Summary

Proteolysis is one of the most important post-translational modifications of the proteome with every protein undergoing proteolysis during its synthesis and maturation and then upon inactivation and degradation. Extracellular proteolysis can either activate or inactivate bioactive molecules regulating physiological and pathological processes. Therefore, it is important to develop non-biased high-content screens capable of identifying the substrates for a specific protease. This characterization can also be useful for identifying the nodes of intersection between a protease and cellular pathways and so aid in the detection of drug targets. Classically, biochemical methods for protease substrate screening only discover what can be cleaved but this is often not what is actually cleaved in vivo. We suggest that biologically relevant protease substrates can be best found by analysis of proteolysis in a living cellular context, starting with a proteome that has never been exposed to the activity of the examined protease. Therefore, protease knockout cells form a convenient and powerful system for these screens.

We describe a method for identification and quantification of shed and secreted cleaved substrates in cell cultures utilizing the cell metabolism as a labelling system. SILAC (stable isotope labelling by amino acids) utilises metabolic incorporation of stable isotope-labelled amino acids into living cells. As a model system to develop this approach, we chose the well-characterised matrix metalloproteinase, MMP-2, because of its importance in tumour metastasis and a large database of MMP substrates with which to benchmark this new approach. However, the concepts can be applied to any extracellular or cell membrane protease. Generating differential metabolically labelled proteomes is one key to the approach; the other is the use of a negative peptide selection procedure to select for cleaved N-termini in the N-terminome. Using proteomes exposed or not to a particular protease enables biologically relevant substrates and their cleavage sites to be identified and quantified by tandem mass spectrometry proteomics and database searching.

Key words: Protease, Proteinases, Matrix metalloproteinase, MMP, Degradomics, Proteomics, SILAC, Shedding, Protease substrate identification, Quantitative proteomics, Tandem mass spectrometry.

Toni M. Antalis and Thomas H. Bugge (eds.), *Methods in Molecular Biology, Proteases and Cancer, Vol. 539*
© Humana Press, a part of Springer Science+Business Media, LLC 2009
DOI: 10.1007/978-1-60327-003-8_8

1. Introduction

Secreted or shed proteins are present at nanomolar concentrations in an abundant background of ECM and serum proteins. Conventional proteomic techniques may be limited by the complexity and broad dynamic range of such samples *(1)*. This chapter describes a metabolic labelling strategy that finds substrates in the extracellular environment of mammalian cells using mass spectrometry-based proteomics. 'Bottom–up' proteomic identification of proteins relies upon analysis of the peptides of a tryptically digested proteome. The more peptides identified per protein increases the confidence that the identification of that protein is correct. However, these analyses are not quantitative. To provide relative quantification the sample origin of the tryptic peptides must be also be identifiable. This can be done post-sample preparation by chemical labelling with an isotopically distinct tag or by metabolic labelling of cells before sample collection. The methodology of proteome labelling for post-sample preparation in order to discover protease substrates has been recently described in detail by us *(2–4)*.

In metabolic labelling SILAC (stable isotope labelling by amino acids) experiments, two cell populations are grown in identical cell culture media supplemented with essential amino acids *(5)*. One cell population is grown with amino acids displaying the natural isotope abundance (light (L) form) whereas the second cell population is grown with selected non-radioactive stable-isotope labelled amino acids (heavy (H) form). Since isotopic amino acids are incorporated into protein in a sequence-specific manner the two proteomes have a residue-specific mass difference that can be easily detected by mass spectrometry analysis. The use of SILAC labelling allows for the discrimination between authentic cellular secreted or shed proteins versus serum proteins and other cell culture medium protein supplements at the mass spectrometry stage. This offers a significant advantage over chemical tagging of proteins or peptides after harvest when serum and medium proteins will also be labelled, and so reducing the specific activity of the labelled cell proteins and complicating data analyses.

When dealing with protease substrate identification in the simplest experimental setup two SILAC isotopes are employed: one form for labelling the 'naive' proteome and another for the same proteome exposed to the protease. Different strategies can be followed for generating biologically relevant proteomes. One approach can be to grow cells expressing the protease of interest in a heavy medium and, on the other side, grow cells expressing an inactive protease mutant in a light medium. A convenient complementary approach examines in vitro protease activity, performed by growing protease null cells separately in heavy and light mediums, and following harvesting of the secretome present in the conditioned media, add recombinant protease to the heavy proteome and inactive or inactivated protease to the light control proteome.

By these two strategies both approaches generate two isotopically labelled proteomes that can be distinguished by mass spectrometry. For cells expressing protease, the two proteomes will differ by those proteins cleaved by the enzyme, such as degradation of secreted proteins and shedding of plasma membrane ectodomains. For processed proteins that are cleaved at only one or a few sites, the problem is more difficult as potentially both cleavage products remain in the system analysis. However, the cell may scavenge cleaved destabilised proteins, thereby leading to differences that can be more easily identified and quantified. This live cell approach also reveals any indirect effects on protein expression resulting from these altered proteins, such as from altered signalling by cleavage of cytokines or their binding proteins. Similarly, studying a protease in a cellular context has the complication, but also power, to analyze at a cell-wide level the effect of the protease on the protease web whether by inactivation of protease inhibitors or activation of protease zymogens. It is clear that upstream and downstream pathways are too simplistic an explanation for understanding proteolysis in a system. Rather we have proposed that proteases interact with each other through a web of interactions. Hence, the best way to study this is in the cellular context, and eventually with technology development, in animal tissues from protease knockout mice or human samples.

For the in vitro class of experiment two types of modification to the proteome will be found. Treated and untreated proteomes will differ in the relative abundance only for those proteins degraded by protease activity, and second, for those substrates processed at one or only a few sites, the generation of these unique cleavage products will only be reflected by the generation of unique neo-peptides in the tryptically digested proteome samples. Hence, the measurement of H/L isotopic ratios can be used for substrate identifications. Some peptides (and hence proteins or their protein domains) have equal H/L isotopic ratios and represent the natural N-terminome that includes proteins with and without a N-terminal methionine or signal peptide, activated zymogens, and basal proteolysis in the sample. At this point in the screen one can decide to ignore the proteins with the same relative abundance (i.e., H/L ratios equal to 1) in the samples and focus the analysis to the rest of the proteome. However, complex samples reduce the coverage of MS analyses and can so make the analysis overly complicated. Therefore, to improve the analysis it is best to utilise a strategy to simplify the peptide mixture selectively by enriching the preparation for peptides of our interest.

A simplification of the mixture can be done removing peptides with unitary H/L ratios from the mixture and thus selectively enrich the proportion of protease-digested peptides. This can be achieved by introducing chemical differences between the N-termini of the (a) proteome representing both the natural and protease-generated products, and (b) internal peptides generated from trypsin digestion. At the protein level, we chemically mask the N-termini of the harvested proteome and after trypsin digestion the tryptic-generated N-termini can be removed using an amine-reactive reagent. The remaining peptides represent the N-terminome including the protease-generated cleavage products. However, even if the number of peptides in our mixture is $\approx 1/10$ of the whole tryptic proteome, its complexity can be further reduced by HPLC fractionating of the sample and using an LC-MS/MS mass spectrometer with a fast duty cycle.

The reduction of mixture complexity results in a simplification of analysis and data interpretation. An additional approach to simplify data interpretation is through an inverse labelling strategy. In contrast to the conventional single experiment approach, a two-labelling experiment approach has a parallel isotopic labelling in which the initial labelling is reversed in the second experiment. Signals from differentially cleaved proteins will distinguish themselves by exhibiting a characteristic pattern of isotope intensity profile reversal. Inverse labelling approach offers reduction in the amount of work spent on data analysis and it eliminates ambiguity in data interpretation. After correction for any systematic error, the analysis of the peptide ratio peak centred around 1.0 provides an estimation of the standard error of the procedure, from which a ratio threshold can be set to determine those peptides that have significantly changed with high confidence and hence protease substrate candidates.

Inverse SILAC labelling of proteolyzed proteome followed by N-terminome enrichment and strong cation exchange (SCX) fractionation and MD-LC MS/MS reduce the complexity of the sample, which (a) reduces the number of MS analyses, (b) improves coverage of the proteome and therefore experimental reproducibility, and (c) allows for the determination of protease-specific cleavage sites. The workflow is summarised in **Fig. 1**. Fully incorporated heavy- and light-labelled secretomes are harvested and concentrated; following the combination of the two-labelled samples, protein is denatured, reduced, and alkylated. The determination of systematic mixing and handling errors is performed by mass spectrometric analysis of a small aliquot (no more than 10 μg is needed) of the trypsin-digested proteome before the enrichment step. Therefore, the H/L ratio standard distribution of each peptidic pair is plotted and used to derive the centre of the curve. These parameters (centre of distribution and standard deviation) are used to correct any systematic error

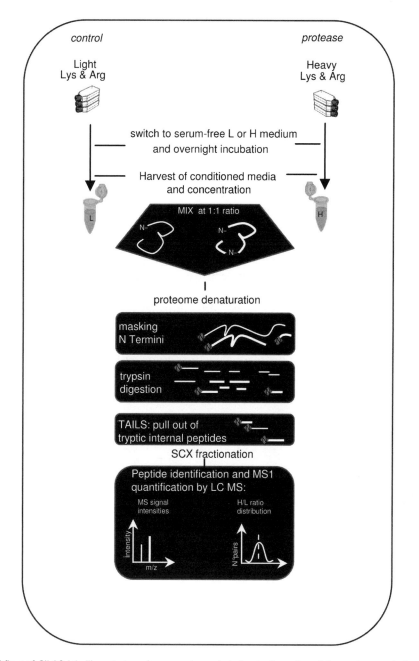

Fig. 1. Workflow of SILAC labelling strategy for screening substrates in the extracellular environment of mammalian cells.

of the ratio quantification of the actual experiment. The rest of the proteome is treated so as to enrich for the protease-cleaved peptides termed 'neo-peptides'. Here, the protein N-termini are blocked by dimethylation using formaldehyde and then digested into peptides with trypsin. These internal peptides are removed by targeting their free N-termini with an amine-reactive polymer.

To improve peptide coverage, the enriched proteome is fractionated by SCX. Samples are analysed by MS/MS to identify peptides and the relative abundance of isotopic peptide pairs from the two samples quantified in MS mode. Further bioinformatics analysis is restricted to only the peptide pairs showing H/L ratios that differ from 1 ± standard deviation. These peptides reveal the proteins affected from the protease activity studied. Finally, by BLAST analysis, the specific cleavage site can be identified.

The protease used for developing this protocol is a member of the matrix metalloproteinases (MMPs), extracellular zinc-dependent endopeptidases implicated in a number of physiological and pathological processes such as cancer and inflammation *(6)*. As a model system, we have used SILAC to identify substrates of MMP-2 from the cell-conditioned medium of cultured human breast cancer cells transfected with MMP-2 compared with vector or an inactive MMP-2 mutant. Similarly, we have identified substrates of MMP-2 in conditioned media of cultured *Mmp2–/–* murine fibroblasts, containing secreted proteins and cytosolic proteins resulting from cell lysis, with the same secretome exposed to human MMP-2 activity. By comparing protease naive proteome with protease-treated proteome the highest signal-to-noise ratio can be obtained for MMP-2 proteolysis.

2. Materials

Warnings are given for hazardous materials, but Material Safety Data Sheets (MSDS) should be read prior to starting the protocol.

2.1. Cell Culture

1. Cell line of choice (adherent cells are usually employed for finding substrate in the extracellular environment).

2. Culture dialyzed medium DMEM (Dulbecco Modified eagle's Medium) with high glucose, L-glutamine, 110 mg/ml sodium pyruvate, and pyridoximetile without L-lysine and L-arginine (Caisson Laboratory, Utah, USA).

3. 'Regular' amino acids (most abundant isotope form): L-arginine monochloride, L-lysine, and L-leucine. SILAC amino acids: L-arg-$^{13}C_6$ monochloride, L-lys-D_4. The heavy isotopic forms of Lys and Arg differ by 6 and 10 Da, respectively, from normal Lys and Arg.

4. Dialyzed fetal bovine serum, supplements, and selection reagents appropriate for cell type used.

5. Sterile phosphate buffered saline (PBS).

6. Sterile-top presterilized vacuum-driven disposable bottle top filter.

7. Dimethyl sulfoxide minimum (DMSO) 99.5% (*see* **Note 1**).

8. ZWITTERGENT 3-14 detergent.

9. Lysis buffer: 50 mM Tris–HCl, 150 mM NaCl, 10 mM EDTA, 0.2% ZWITTERGENT, 10 mM EDTA, 1 mM phenylmethyl-sulfonyl fluoride (PMSF); pH 8.0 (*see* **Note 1**).

2.2. Sample Collection and Preparation

Solvents should be of the highest available grade. Mass spectrometry laboratories routinely use Eppendorf® microcentrifuge tubes (Eppendorf AG, Germany). Do not store organic solvent in microcentrifuge tubes that are specially coated to reduce protein binding as the coating can be solubilized and will lead to chemical contaminations detectable by MS. Wear always nitrile (not latex) gloves. Use high-quality deionised H_2O throughout such as Milli-Q® or Barnstead Nanopure processed H_2O.

2.2.1. Concentration of Conditioned Medium

1. 0.5 mM PMSF (*see* **Note 1**).

2. 1 mM EDTA ethylenediaminetetraacetic acid.

3. 10 mM leupeptin (*see* **Note 1**).

4. 10 mM pepstatin A.

5. Versine: 140 mM NaCl, 2.7 mM KCl, 10 mM Na_2HPO_4, 1.8 mM NaH_2PO_4, 0.5 mM EDTA, 1 mM glucose; pH 7.4.

6. Filtration unit (0.2-μm pore size).

7. Concentrators with 5-kDa cut-off membrane, Amicon (Millipore™, California, USA), 0.05 M NaOH (wash Amicon concentrator membrane for improving recovery; *see* **Note 2**).

8. 50 mM Hepes (4-(2-hydroxyethyl)-1-piperazineethanesulfonic acid), pH 7.2.

2.2.2. Cell Counting and Protein Quantification

1. Trypan blue solution 0.4%.

2. Hematocytometer.

3. Bicinchoninic Acid (BCA)™ Protein Assay kit Pierce (Thermo Fisher Scientific, Inc., MA, USA).

4. Spectrophotometer reader for protein concentration assay.

5. Bradford assay reagents.

2.3. Proteome Denaturation, Reduction, and Alkylation at the Protein Level

1. Use high-quality deionised H_2O throughout such as Milli-Q® or Barnstead Nanopure.

2. Employ pH paper indicator strips to monitor the pH along the procedure.

3. Detergent, Rapigest™ (Waters Corporation, Massachusetts, USA).

4. Denaturation buffer: 0.2% Rapigest™, 150 mM HEPES; pH 7.4.

5. Reducing reagent, 7 mM TCEP (2-carboxy-ethyl phosphine hydrochloride) Tris (*see* **Note 1**).

6. Alkylating agent, iodoacetamide (*see* **Note 1**).

7. Cyanoborohydride, ALD coupling solution (Sterogene Bioseparation, Inc., Carlsbad, CA) (*see* **Note 1**).

8. Formaldehyde 37% (w/w) solution (*see* **Note 1**).

9. Primary amine blocking reagent: Ammonium chloride or ammonium bicarbonate (*see* **Note 3**).

2.4. Sample Cleanup and Generation of Tryptic Peptides

1. Acetone (*see* **Note 1**).

2. Methanol (*see* **Note 1**).

3. Mass spectrometry grade trypsin (Promega Corporation, WI, USA).

4. Reagents for silver staining of SDS-PAGE gel.

2.5. Enrichment of Tryptic Peptides of Interest (Polymer Pullout)

1. In-house polyaldehyde polymer (soon to be available comercially).

2. Concentrators with 5-kDa cut-off membrane, Microcon (Millipore™, California, USA). Wash the filter membrane to reduce losses (*see* **Note 2**).

3. Slide-A-Lyzer 10 K MWCO dialysis cassettes (0.5–3 ml) (Pierce, Thermo Fisher Scientific, Inc., MA, USA).

4. Ammonium bicarbonate.

5. Benchtop microcentrifuge.

2.6. Fractionation and Preparing Samples for LC-MS Analysis

1. Material for casting STAGE C18-SCX-C18 tips: high-performance extraction discs, C18 and strong cationic exchange SCX 3M Empore (Empore™ Products, Minnesota, USA).

2. PFPA: Perfluoro pentanoic acid nanofluoropentanoic acid (*see* **Note 1**).

3. Ammonium Acetate (AcONH$_4$).

4. Trifluoroacetic acid (TFA) (*see* **Note 1**).

5. Vacuum evaporator centrifuge.

6. Acetonitrile (*see* **Note 1**).

7. Acetic acid (*see* **Note 1**).

8. 99% Pure ethanol (*see* **Note 1**).

9. C18-SCX-C18 StageTips (16-gauge, P200 pipette) prepared as described in (*11*).

10. One 96-well plate for LC-MS injection.

11. Buffer A: 0.1% TFA, 5% acetonitrile.

12. Buffer B: 0.1% TFA, 80% acetonitrile.

13. 500 mM AcONH$_4$.

14. 20 mM AcONH$_4$, 0.1% v/v PFPA, 15% acetonitrile.

15. 50 mM AcONH$_4$, 0.1% v/v PFPA, 15% acetonitrile.

16. 100 mM AcONH$_4$, 0.1% v/v PFPA, 15% acetonitrile.

17. 500 mM AcONH$_4$, 0.1% v/v PFPA, 15% acetonitrile.

2.7. Mass Spectrometry Analysis

1. Liquid chromatography LC-tandem mass spectrometer.

2. Protein and peptide software tools: Mascot (Matrix Science: http://www.matrixscience.com), Xcalibur™, DTA-Super-charge (CEBI, http://msquant.sourceforge.net), and MultiRawPrepare (http://www.cebi.sdu.dk/software).

3. Quantification software tools (such as MS Quant): http://msquant.sourceforge.net/.

4. Excel (from Microsoft Office) for extracting H/L ratios from MSQuant.

5. MATLAB® software for the calculation of Gaussian parameters of H/L distribution (centre of the distribution and standard deviation).

6. BLAST (http://www.ncbi.nlm.nih.gov/blast/Blast.cgi) is the search tool used for the identification of substrate and cleavage sites.

3. Methods

3.1. Prior Preparation

Ab initio, it is important to determine the number of cells required to detect significant signal for MS analysis. For that purpose, do not use costly SILAC medium.

3.2. Designing Experiments

Most experimental designs compare multiple biological conditions analyzing two proteomes at the time. When trypsin is used as endopeptidase to digest proteome into peptides (*see* **Note 4**), the use of both heavy lysine and arginine allows quantification of every peptide pair (except the carboxyl-terminal peptide of the proteins) because trypsin specifically cleaves C-terminal to these residues. If the mixture is not too complex, it will be also beneficial to compare multiple conditions within the same mass spectrometry analyses. For instance, by using three isotopically distinct forms of arginine and lysine it is possible to compare three cell populations in a single experiment. When designing the experiment bear in mind that the number of peptides in the mixture will be directly proportional to the number of distinct isotopic acids employed, so increasing complexity.

3.3. Preparation of SILAC Media

1. Prepare concentrated amino acid stock solutions (1,000×) to reconstitute SILAC DMEM light and heavy media (*see* **Note 5**). Since SILAC medium lacks three essential amino acids, these three need to be added. For metabolic labelling, one SILAC amino acid is sufficient to distinguish the two proteomes. However, employing SILAC Lys and Arg results in better sensitivity and sequence coverage since all tryptic peptides will be labelled (*see* also **Subheading 3.2**).

2. Supplement SILAC dialysed medium with 'regular' (most common isotope) leucine. Add dialysed serum, supplements, and selections appropriate for cell type to the SILAC depleted medium. In similar fashion, prepare also the serum-free medium.

3. Mix well and divide into two equal volumes. Add light isotopes of arginine and lysine to one-half and the heavy isotopes to the other half (*see* **Note 6**). To prepare H and L medium for growing cells, perform the same procedures of the serum-free media.

4. Filter the four bottles of reconstituted media through a 0.2-µm filter into tissue-culture grade filter bottles. Store media at 4°C protected from the light for up to 3 months.

3.4. To Determine the Extent of Metabolic Labelling with SILAC Amino Acids

To accurately compare the relative abundance of proteins from two proteomes with SILAC, full incorporation of SILAC amino acids is necessary (higher than 98%). Complete incorporation is usually achieved within 6–7 passages of the cells in the medium containing the isotope-labelled amino acids.

1. Grow cells in heavy SILAC medium for 6–7 cell doubling times (*see* **Note 7**). Verify that the cells do not change morphology.

2. Wash the cells gently three times with 20 ml of PBS and incubate overnight in heavy serum-free medium (*see* **Note 8**).

3. Detach the cells grown in heavy medium using versine or trypsin. Harvest at least 1×10^6 cells to test the efficiency of SILAC incorporation.

4. Count live cells with a hemocytometer chamber using the trypan blue dye exclusion method. Verify that heavy and light cells are of the same number.

5. Centrifuge the cells at $500 \times g$ for 5 min and resuspend in 0.5 ml of pre-chilled (4°C) lysis buffer and incubate for 1 h on ice.

6. Centrifuge cell lysate at $1,500 \times g$ for 20 min at 4°C.

7. Determine protein concentration of supernatant by Bradford assay.

8. Mix heavy and light lysates in 1:1 ratio.

9. Fractionate proteome by 1D SDS-PAGE. Load 70–90 µg of proteome per lane. Electrophorese heavy and light proteomes in one lane and heavy proteome alone in a second lane.

10. After staining the gel with Coomassie Blue R250, excise several protein bands per lane.

11. Perform trypsin in-gel digestion (*see* **Subheading 3.5**).

12. Analyze digested gel samples according to standard techniques commonly used in mass spectrometry proteomics. The degree of incorporation is determined by manually inspecting the MS spectra *(16)*. The labelling yield must be greater than 90%. Identify peptides and proteins using database search software, making sure to add modified masses of SILAC amino acids to the list of possible modifications in search engine. In the case of heavy proteome data look for the presence of light peptides making sure that peptides containing the SILAC amino acids are present only in the heavy form (**Fig. 2**).

13. Once the number of cell divisions needed to achieve 100% incorporation has been established, grow the cells separately in heavy and light SILAC media and freeze heavy and light cells to form the stock cells to be used for the actual experiments.

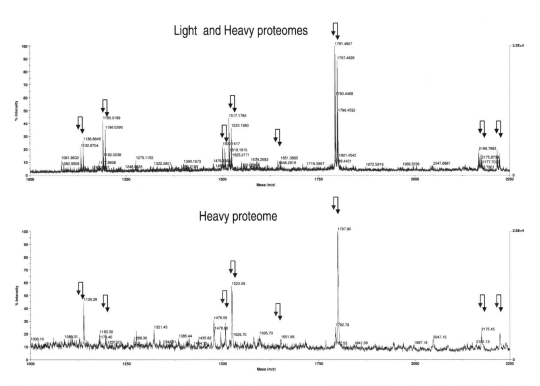

Fig. 2. Validation of heavy SILAC isotopic incorporation. In the *upper* panel MALDI MS spectrum of tryptic digestion of combine (H + L) proteome. In the *lower* panel MALDI MS spectrum of tryptic peptides from the digestion of H proteome. *Double arrows* indicate the *m/z* values where SILAC isotopic peak pair is expected to be.

3.5. Trypsin In-Gel Digestion

To check SILAC amino acid incorporation it is not important to obtain maximum peptide coverage from the digestion. Therefore, the following protocol describes a quick way of performing trypsin in-gel digestion. It can be improved if desired by introducing in-gel reduction and in-gel alkylation of proteins.

Unless otherwise noted, the various steps of the procedure must be performed at room temperature and all incubation steps under shaking conditions.

1. Rinse the stained gel in HPLC grade water for 10 min.

2. Excise the band of interest and cut into small cubes (1 mm × 1 mm). Transfer the gel pieces into a clean 0.5-ml polyethylene sample vial.

3. Wash with deionised 0.2 ml of nanopure water.

4. Add 50-μl HPLC grade acetonitrile to dehydrate the gel pieces; after 15 min, discard the supernatant.

5. Repeat **steps 3** and **4** until the gel pieces are transparent.

6. Add enough ice-cold digestion solution (trypsin 10 ng/μl in 25 mM ammonium bicarbonate, pH 8.0).

7. Incubate on ice for 20 min to allow trypsin to penetrate gel pieces (the cold temperature helps to prevent autolysis of trypsin).

8. Replace the trypsin solution with 25 mM ammonium bicarbonate, pH 8.0 buffer.

9. Incubate at 37°C overnight.

10. Extract peptides by adding 50 μl of acetonitrile, incubate for 15 min in sonication bath at 37°C, and collect the supernatant in a clean microfuge tube. Peptide extraction is more efficient by sonication rather than shaking the tubes.

11. Hydrate gel cubes by adding 50 μl of 5% (v/v) formic acid to the gel pieces and incubate at 37°C for 15 min.

12. The second extraction of peptides is performed by adding 50 μl of acetonitrile and incubating for 15 min in a sonication bath at 37°C. Collect the supernatant and combine it with the first peptide extraction.

13. Concentrate the supernatant using a centrifugal vacuum concentrator (30°C) to 5 μl. Do not allow the sample to dry out.

3.6. Preparation of Heavy and Light Proteomes

3.6.1. Metabolic Labelling

After verifying 100% incorporation of heavy amino acids, quantitative experiments between two types of cell can be performed (*see* **Subheading 3.4**).

1. To begin the labelling process, split a confluent flask into two separate flasks. Grow one culture with light SILAC supplemented medium and the other with heavy SILAC supplemented medium (*see* **Note 9**).

2. Passage cells as needed, taking care to keep cells in their respective SILAC media using the standard cell culture technique. Make sure that the number of heavy and light cells is the same for all passages. Grow cells in culture medium in roller bottles (*see* **Note 10**) or T-175 tissue culture flasks for at least seven doubling times to allow complete incorporation of SILAC amino acids.

3. Grow sufficient numbers of light- and heavy-labelled cells (*see* **Note 11** and **Subheading 3.1**).

4. When the cells are 80–90% confluent at the time of medium collection (*see* **Note 12**), wash the cells three times with PBS (*see* **Note 8**).

5. Switch to the corresponding light or heavy media without serum and incubate for the desired time. Perform a cell treatment if desired (*see* **Note 13**).

3.6.2. Sample Collection and Preparation

Conditioned Medium

1. Harvest heavy and light conditioned media separately and immediately add protease inhibitors (final concentrations 0.5 mM PMSF, 1 mM EDTA, 10 μM leupeptin, 10 μM pepstatin A) (*see* **Note 14**). Keep the conditioned media cool, at 4°C, or on ice from this point on.

2. Clarify conditioned medium by centrifugation to remove any cells ($500 \times g$, 5 min) and filter through 0.2-μm pore size to remove particulate matter (*see* **Note 15**).

Cell Counts

1. Harvest light and heavy cells separately, detaching the cells with versine or trypsin.

2. Check whether the cell conditions were similar in the two cultures, making sure that the heavy and light cultures have the same number of cells. Determine cell viability using hemocytometer with the trypan blue dye exclusion method.

Concentration of Conditioned Medium

1. Concentrate filtered conditioned media using 5-kDa cut-off concentrators (Centriprep and Microcon, Amicon) $4,000 \times g$ until the volume goes down to a few millilitres (*see* **Note 2**).

2. Exchange the buffer of collected medium washing with 50 mM HEPES using three times the volume of collected conditioned medium (*see* **Note 16**). Stop the centrifugation of Amicon filters when the retentate volume is about 1 ml.

Protein Quantification

Determine the protein concentration of concentrated conditioned medium using the micro-scale protocol supplied with both the BCA protein assay kit and Bradford assay. The BCA protein assay is the benchmark for accurately measuring low level of protein concentration; however, this assay may be affected by other substances occasionally present in concentrated conditioned medium, including detergents, lipids, buffers, and reducing agents. This obviously requires that the assays also include a series of standard solutions, each with a different, known concentration of protein, but otherwise having the same composition as the sample solutions. The Bradford assay is faster, and gives a more stable colorimetric response than the assay described earlier. Like the other assays, however, its response is prone to influence from non-protein sources, particularly detergents, and becomes progressively more non-linear at the high end of its useful protein concentration range.

Mixing Heavy and Light Proteomes

Pool concentrated conditioned media from heavy and light proteomes in a 1:1 ratio.

3.7. Preparation of Proteome for MS Analyses: From Proteins to Peptides

The proteins are denatured with 0.2% Rapigest™ detergent and reduced in 7 nM TCEP to expose the cysteine residues for alkylation. For some of following steps, measuring the pH with pH paper is required to minimize sample losses (if it is not differently indicated adjust it with 0.1 M NaOH and 0.1 M HCl).

3.7.1. Protein Denaturation, Reduction, and Alkylation

1. Resuspend pellet in 80-μl denaturing buffer, add concentrated Hepes buffer, pH 7.2, so that the final molarity is 250 mM, and vortex briefly. Verify that the pH of reconstituted buffer is as expected (i.e., pH 7.2).

2. Denaturation: Add 0.2% Rapigest™ and incubate at 85°C for 10 min. Verify that pH has not changed.

3. Reduction: Add 20× reducing reagent (TCEP) solution and incubate at 65°C for 30 min. Check that the pH stayed slightly higher than 7.0.

4. Alkylation: Following cooling of the sample on ice, verify that the pH has not changed. Start the alkylation reaction by incubating 20 mM iodoacetamide in the dark at room temperature. The reaction is complete in 1 h. Verify that there was no pH variation.

3.7.2. Determination of Systematic Error and Procedure Accuracy

Since we will find substrates among the proteins not equally present in the two isotopic proteomes, it is extremely important to correct any eventual systematic mixing error. The 1:1 mixing ratio between isotopic proteomes is checked along the procedure by counting cells and by protein concentration assays; however, the accuracy of these assays is far above MS resolution (*see* **Subheadings "Cell Counts" and "Protein Quantification"**). Nevertheless, an eventual systematic error can be easily identified and corrected if statistically characterized. One way to determine the occurrence of any systematic error can be performed by plotting the peptidic H/L ratio standard distribution of a small sample of crude proteome (around 10 μg) harvested before the sample pull-out step (*see* **Subheading 3.7.6**). If the centre of the ratio distribution of experimental data is shifted, it can justifiably be normalised, centering the distribution on 1. On the other side, the accuracy of each data set can be validated looking at the average standard deviation of the H/L distribution (the error of the procedure usually is never intrinsically lower than 0.2).

3.7.3. N-Terminal Blocking of Proteins

1. Block the free primary amines of proteins incubating the sample with 20 mM ALD reagent (Cyanogen Bromide (CNBr)) and 40 mM formaldehyde at 37°C for 1 h (the acetylation reaction can be extended up to 18 h) (*see* **Note 3**).

2. Stop the reaction quenching with 20 mM ammonium chloride (NH$_4$Cl) at 37°C for 30 min (*see* **Note 3**).

3.7.4. Sample Cleanup by Acetone Precipitation

1. Reduce sample volume using a speed vac to improve and facilitate the precipitation process.

2. Add 8 volumes of pre-chilled (–70°C) acetone plus 1 volume of pre-chilled methanol to 1 volume of the sample (at least 60 μg of sample on ice; vortex and incubate for 2 h at –70°C to precipitate proteins).

3. Centrifuge (13,000 × *g*, 15 min) at 4°C (a small pellet should be visible). Carefully pour off supernatant and allow the pellet to air-dry until just moist but not powder dry (*see* **Note 17**).

3.7.5. Trypsin Digestion

1. Resuspend the pellet using 20 μl of 100 mM NaOH shaking for 3 min, then add ~20 μl of 1 M Hepes, pH 8. Shake for additional 3 min, and add double-distilled water to bring the protein concentration to 1 mg/ml.

2. Since catalytic efficiency of trypsin is optimal at pH 8, ensure that the pH is close to this.

3. For a complete resuspension of proteome, shake the sample at 37°C for 15 min.

4. Collect a volume that contains 6 μg of proteome and transfer it in a different tube (it will be used to check the completion of trypsin digestion).

5. Add MS spectrometry grade trypsin in the trypsin:proteome ratio of 1:200 (w/w).

6. Incubate at 37°C for 18 h.

7. Add same amount of trypsin and let the digestion go for additional 4 h.

8. To check the completion of trypsin digestion, run an SDS-PAGE gel (12% acrylamide) of 6 μg of proteome before trypsin digestion (**step 4**) and the same volume of proteome after the digestion. Since tryptic peptides run off the gel, after silver staining the gel, only the lane loaded with the predigested proteome should be visible (*see* **Note 18**).

3.7.6. Reduction of Proteome Complexity

To simplify peptide mixtures, internal tryptic peptides can be removed and so selectively increase the proportion of peptides that have been cleaved by the studied protease. Owing to similar chemistry of the primary amine groups of the N-terminus of a protein and of the Lys side chains, the amines are chemically blocked and then the sample is trypsinised. Except for the N-termini of the proteins and the protease-cleavage products, which are blocked, only the internal tryptic peptides now have a free amino group. This difference can be used to allow for internal tryptic peptides pulled out using an amine-reactive reagent that captures internal peptides from the mixture. The remaining peptides represent the N-terminome of the sample and include the protease-generated N-termini.

Several strategies have been developed to selectively analyse free N-terminal peptides, including specific N-terminal sequencing of gel separated and blotted proteins (7), selective modification of N-terminal serine or threonine residues (8), and selective modification of the hydrophobicity of peptide mixture to preferentially expose N-terminal peptides by diagonal chromatography (9). Another strategy can be to use N-hydroxysuccinimide (NHS) ester derivative of biotin and then remove the biotinylated peptide by passing the mixture over streptavidin. For this purpose functionalized beads as cyano bromide (CNBr)-activated Sepharose can be employed (10). In contrast, we use an in-house made aldehyde-functionalized polymer that is CNBr activated and incubated with the tryptic proteome at 37°C overnight. This will be commercially available in 2009.

3.7.7. Proteome Fractionation

Tandem mass spectrometry is a powerful technique for improving peptide identification of complex proteome by two consecutive stages of mass spectrometric scan (i.e., MS1 and MS/MS). However, without a more favourable strategy, the identification of nanomolar substrates can be remote due to the abundance of structural proteins of the extracellular matrix proteins or other prominent proteins in the sample, including any serum proteins. Furthermore, as the SILAC label strategy introduces a mass shift detectable at the MS1 level, the SILAC labelling of the proteome doubles the number of peptides to be analysed. This complexity can be reduced by fractionating the mixture leading to higher coverage for peptidic identifications. Several forms of fractionation have been developed such as liquid chromatography (11–14) or isoelectric focusing (7).

Here, we will describe a cheap, quick, yet very effective method (16) perfectly suitable for medium-scale experiments (when only few micrograms of proteome can be employed) (see **Note 19**). We use a previously described pipet tip-based peptide multidimensional fractionation system named StageTips (STop and Go Extraction Tips) (16). This solid phase extraction technique uses three very small disks of membrane-embedded separation. C18 and SCX materials are stacked within the pipet tip according to the following order C18–SCX–C18. In this way, the resulting fractionations can be as well desalted, filtrated, and concentrated, being thus ready for mass spectrometry run (16).

The fractionation can be performed as follows. For all the following steps, load the indicated volumes from the top of the STAGE tip C18–SCX–C18, and press the liquid through STAGE tip only once.

Conditioning steps: To activate the tip C18 and SCX materials, wet the tip loading 20 µl of the following substances and discard the solvent.

1. Methanol.

2. Buffer B.

3. Buffer A.

4. 500 mM AcONH$_4$.

5. Buffer A.

Sample loading to C18–SCX–C18 StageTip:

6. Sample loading (pre-mix the sample and buffer A in the ratio 1:3).

7. Buffer A 20 µl.

Fractionation to 96-well injection plate:

8. Buffer B 20 µl – collect as flow through fraction 0.

9. 20 mM AcONH$_4$, 0.1% PFPA, 15% acetonitrile 50 µl.

10. Buffer A 20 µl.

11. Buffer B 20 µl – collect as fraction 1.

12. 50 mM AcONH$_4$, 0.1% PFPA, 15% acetonitrile 50 µl.

13. Buffer A 20 µl.

14. Buffer B 20 µl – collect as fraction 2.

15. 100 mM AcONH$_4$, 0.1% PFPA, 15% acetonitrile 50 µl.

16. Buffer A 20 µl.

17. Buffer B 20 µl – collect as fraction 3.

18. 500 mM AcONH$_4$, 0.1% PFPA, 15% acetonitrile 50 µl.

19. Buffer A 20 µl.

20. Buffer B 20 µl – collect as fraction 4.

Sample prep for LC/MS/MS:

21. Evaporate solvents in the 96-well injection plate using SpeedVac

22. Add sample buffer or buffer A 3 µl.

3.8. MS Data Acquisition

To maximize the number of analyzed peptides the use of liquid chromatography (LC) mass spectrometer is recommended. Analysis of the fractions can be performed using a linear trapping quadrupole-Fourier transform mass Spectrometer (LTQ-FT, Thermo Electron, Bremen, Germany), LTQ-Orbitrap (Thermoelectron), or similar instruments.

3.9. Data Analysis

3.9.1. Conversion of Raw Files into Mascot Generic Format (mgf):

1. The peak list can be extracted from raw data files to mascot generic format (mgf) using Extract_MSN.exe (ThermoFisher Scientific).

2. Centroided peaks and charge states are corrected using DTA_Supercharge (CEBI, http://msquant.sourceforge.net).

3. The generated five peak list files (.mgf files) can be concatenated to generate a single large peak list file using Multi-RawPrepare software (http://www.cebi.sdu.dk/software).

3.9.2. Database Search

The resulting mgf file is used to identify peptide sequences using the MS database search algorithm MASCOT (Matrix Science, London UK). The parameter criteria are as follows: semi-trypsin cleavage specificity (only after Arg) with up to two missed cleavages, cysteine carbamidomethyl fixed modification; Arginine^{13}C$_6$, Lysine (D4), di-methylation (K), di-methylation (N-Term), oxidation (M) variable modifications; peptide mass tolerance ± 10 ppm; fragment mass tolerance ± 0.6 Da, and scoring scheme ESI-TRAP.

3.9.3. Quantification

The ratio signal intensities from peptide pair indicate the relative abundance of that specific peptide from one condition to the other, allowing the identification of proteins related to the protease activity.

1. Before starting quantification the data must be manually inspected *(16)*. Verify that identified peptides are separated by the expected mass difference. The MS/MS spectra from the pair support the peptide identification.

2. MSQuant (http://msquant.sourceforge.net) is used for parsing Mascot files, iterative mass recalibration, and determination of H/L ratios. After the recalibration only peptides with error lower than 5 ppm are considered for the quantification of isotopic peptide ratio. The heavy over light ratio (H/L) is calculated per each pair of peptide, and the value is extracted into Excel files. Average, standard deviation, and centre of the H/L distribution are calculated using Matlab software package (Mathworks, Inc.).

3.9.4. Inverse Labelling

Two parallel inverse labelling experiments are performed where the labelling is reversed in the second experiment (i.e., in the first experiment the control proteome is light isotope labelled and the protease-treated proteome is heavy labelled; in the second experiment the isotopes employed are the opposite). A rapid identification of protease-affected proteins is achieved by plotting the histogram distribution of H/L peptidic ratios (*see* **Fig. 3**). As expected the distributions of the two swapped experiments are oppositely shifted with respect to 1 (i.e., similar abundance of light and heavy isotopic signals in both experiments). Thus, the

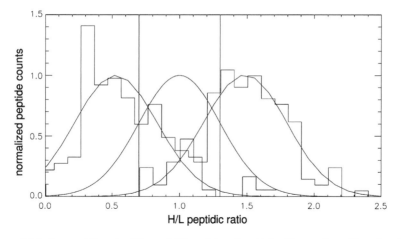

Fig. 3. Histogram distributions and Gaussian fittings of *H/L* peptidic ratios obtained from cell-conditioned media of cultured murine fibroblasts. On the *left*, the direct experiment *H/L* distribution is reported where the protease-treated proteome is heavy isotope labelled. On the *right*, the reverse experiment is reported where the control (inactive-protease-treated) proteome is heavy isotope labelled. In the *middle*, the Gaussian fitting is used for the determination of standard deviation and procedure accuracy (*see* Subheading 3.7.2).

proteins that are significantly influenced by protease activity show H/L ratios lower than 0.7 in the first experiment and greater than 1.3 in the swapped counterpart experiment (*see* **Fig. 3**).

By applying the inverse labelling strategy, the effort spent in analysing irrelevant peptides with no differential change is eliminated (i.e., pair matching 1 intensity ratio). The advantages offered by this approach are obvious. When responding to a perturbation, if a protein is from serum leftover contamination that does not consistently occur in the swapped experiment, its peptide will have no isotopic counterpart in the analysis using the single-experiment approach. When the labelling is reversed between the two inverse labelling experiments, the isotope swap will be clearly detected on these peptides. Thus, the inverse labelling pattern is unique to proteins with real changes at both quantitative and qualitative levels. Although an additional experiment must be performed, in our opinion, this procedure tackles three problems: (a) data reduction of irrelevant signals, (b) quick focus on signal of interest only, and (c) conclusions of minimal ambiguity.

3.9.5. Identification

Protein BLAST (http://www.ncbi.nlm.nih.gov/blast/Blast.cgi) algorithm is used for visualizing the position of peptides within the protein sequence and hence the cleavage site.

3.10. Validation:
Deciding which 'Hits'
Are Significant

Data analysis and validation of the results from MS/MS searches have become major issues of mass spectrometry-based proteomics. To improve reliability of the results some suggestions follow.

1. Validate peptide and protein identification using standard proteomics guidelines *(17)*; SILAC amino acid residue constraints for peptide identification should be applied.

2. Use peptide identification and retention time information together to locate a specific peptide within the chromatographic run.

3. Calculate the mass difference between the light and heavy peptides in the SILAC pairs; this should match the peptide identification and the number of SILAC amino acids.

4. Evaluate the standard error from the isotopic ratio distributions. Identify all known substrates for the protease under study and set 'cutoffs' for test:control ratios based on these. This will be more biologically relevant for the system under study than setting arbitrary numerical. Select proteins with ratios above the highest cutoff and below the lowest cutoff.

5. Consider that a contamination of authentic secreted protein by proteins released in the medium by cell lysis occurs even in the best cultures. Choose those proteins that may be biological substrates of the protease under study (based on location, function, etc.) for biochemical validation.

Even if innovative proteomic approach allows for screening low abundant substrates within a very complex protein mixture, classical methods remain still suitable for validating candidate substrates (i.e., in vitro cleavage using purified protease and putative substrate, follow-up with cell-based and in vivo models).

4. Notes

1. *Danger*: The reagent may be fatal if inhaled, absorbed through skin, or swallowed. This reagent should be handled in a fume hood, and protective eyewear, glasses, and gloves should be worn.

2. To reduce losses of proteins in adsorption to the filter, rinse centricons with NaOH 0.05 M, centrifuge for 5 min at $1,000 \times g$, and wash with deionised water before loading conditioned medium.

3. The acetylation reaction should be handled carefully, preferably in a fume hood. Do not breathe the acetylation reaction vapours.

4. Since trypsin is a very robust enzyme this is the most employed protease useful in preparing samples for mass spectrometry analyses; however, there are also a variety of other specific endoproteases that can be used in proteomics such as chymotrypsin, pepsin, GluC, and V8. If it is suspected that the studied enzyme cleaves after Arg the protocol can still be used but do not use trypsin as the endoprotease for peptide preparation.

5. All reagents and equipment coming in contact with live cells must be sterile. Medium that is deficient in multiple amino acids should always be reconstituted to its original basal formulation as prescribed by manufacture's instructions. 1,000× stocks of amino acid are recommended. However, to reduce losses of costly SILAC amino acids it is better to filter these once inside the reconstituted medium (this step is unnecessary if all amino acid stocks are already sterile filtered).

6. It could also be possible to use only one isotopic amino acid. However, employing double labelling results in better sensitivity and sequence coverage (*see* also **Subheadings 3.2** and **3.3, step 1**).

7. Standard sterile tissue culture technique should be used. All culture incubations are performed in a humidified 37°C, 5% CO_2 incubator.

8. These washes remove abundant serum proteins, which would mask low abundant proteins secreted from cultured cells. In case cells start detaching use room-temperature PBS or SILAC-depleted medium.

9. Split the cells into low-density cultures of about 10–20% confluence, depending on the doubling rate of cells. To save SILAC media, cells can be cultured in small dishes and expanded after five doublings.

10. Roller bottles may be useful as they allow a small amount of medium (50 ml) to bathe several fold more cells than in a flask. However, some cell lines do not adhere well; in this case, adding 50 mM Hepes to culturing media might improve cell adhesion.

11. The number of cells required depends on the type of experiment to be performed. For protease substrate identification at least 100 μg of proteome needs to be used as starting material. Usually, conditioned medium from 4×10^7 cells (3 T-175 flasks for each heavy and light SILAC cells) should be sufficient to perform three technical replicates.

12. Do not let cells reach 100% confluency as uncharacterized signalling events may occur, such as receptor internalization.

13. Pre-filtered solutions of recombinant proteases or inhibitors could be exogenously added to cell cultures.

14. Protease inhibitors are required to 'freeze' the profile of shed proteins upon termination of the experiment and to prevent further proteolytic degradation, for example. A specific inhibitor of the protease of interest could also be added at this point. Care should be taken that inhibitors of trypsin are removed prior to trypsin digestion. Amicon concentration step guarantees the removal of small molecules (*see* **Note 13**).

15. Samples can be stored at –80°C for 1 week; however, precipitation of most abundant proteins may occur. For best results, the samples should be concentrated directly after clarification.

16. A disadvantage of buffer exchange is that concentrating also facilitates removal of small molecules (such as amino acids, phenol red, and protease inhibitors) that would interfere with the next steps. Do not use an amine-containing buffer (e.g., Tris or ammonium bicarbonate) as free amines will prevent the reactions of the next steps. *Disadvantage*: Concentrating large volumes is relatively time-consuming. There are various options for concentrating conditioned medium. The C4 and C18 hydrophobic resins allow quick and easy concentration of large samples; however, they cannot be employed if the conditioned media contains phenol red. Ammonium sulphate precipitation may give variable recovery depending on protein concentration. Acetone and TCA precipitation should not be used as precipitation of small molecules interferes with the determination of protein concentration by BCA assay and with labelling steps.

17. Acetone precipitation improves trypsin digestion of proteome because it removes chemical substances that interfere with trypsin activity and renders the proteins more susceptible to degradation.

18. To obtain good MS data it is very important to digest the entire proteome. If trypsin digestion is not complete, make sure that the pH is 8 and add an additional aliquot of trypsin (ratio 1:200 w/w) and incubate for additional 18 h.

19. After the polymer pullout step the proteome amount is 1/10th of the starting material. So, starting with 60–100 µg of proteome it ends up having no more than 6–10 µg for mass spectrometry analysis. This amount is totally suitable for obtaining five fractions from C18-SCX-C18 STAGE tip.

Acknowledgments

Funding for this work was from the National Cancer Institute of Canada (NCIC) and the Canadian Institutes of Health Research (CIHR).

References

1. Guo, L., Eisenman, J.R., Mahimkar, R.M., Peschon, J.J., Paxton R.J., Black, R.A. and Johnson, R.S. (2002) A proteomic approach for the identification of cell-surface proteins shed by metalloproteases. *Mol. Cell Proteomics* 1, 30–36.

2. Butler, G.S., Dean, R.A., Smith, D. and Overall, C.M. (2008) Membrane protease degradomics: proteomic identification and quantification of cell surface protease substrates. In: Peirce, M. and Wait, R. (eds.). *Proteomic Analysis of Membrane Proteins: Methods and Protocols*, Humana, Ottowa, NJ, in press.

3 Dean, R.A., Smith D. and Overall, C.M. (2007). Proteomic identification of cellular protease substrates using isobaric tags for relative and absolute quantification (iTRAQ). *Curr Protocols Protein Sci.* Supplement 49, 21.18.1-21.18.12

4. Butler, G.S., Dean, R.A., Morrison, C.J. and Overall, C.M. (2008). Identification of cellular MMP substrates using quantitative proteomics: isotope-coded affinity tags (ICAT) and isobaric tags for relative and absolute quantification (iTRAQ). In: Clark, I. (ed.). *Methods in Molecular Biology*, Humana, Totowa, NJ, in press.

5. Ong, S.E., Blagoev, B., Kratchmarova, I., Kristensen, D.B., Steen, H., Pandey, A. and Mann, M. (2002) Stable isotope labeling by amino acids in cell culture, SILAC, as a simple and accurate approach to expression proteomics. *Mol. Cell Proteomics* 1(5), 376–386.

6. Sternlicht, M.D. and Werb, Z. (2001) How matrix metalloproteinases regulate cell behavior. *Annu. Rev. Cell Dev. Biol.* 17, 463–516.

7. Yamaguchi, M., Nakazawa, T., Kuyama, H., Obama, T., Ando, E., Okamura, T., Ueyama, N. and Norioka, S. (2005) High-throughput method for N-terminal sequencing of proteins by MALDI mass spectrometry. *Anal. Chem.* 77, 645–651.

8. Chelius, D. and Shaler, T.A. (2003) Capture of peptides with N-terminal serine and threonine: a sequence-specific chemical method for peptide mixture simplification. *Bioconjg. Chem.* 14, 205–211.

9. Gevaert, K., Van Damme, P., Martens, L. and Vandekerckhove, J. (2005) Diagonal reverse-phase chromatography applications in peptide-centric proteomics: ahead of catalogue-omics? *Anal. Biochem.* 345, 18–29.

10. Akiyama, T.H., Sasagawa, T., Suzuki, M. and Titani, K. (1994) A method for selective isolation of the amino-terminal peptide from alpha-amino-blocked proteins. *Anal. Biochem.* 222, 210–216.

11. Washburn, M.P., Wolters, D. and Yates, J.R., III (2001) Large-scale analysis of the yeast proteome by multidimensional protein identification technology. *Nat. Biotechnol.* 19, 242–247.

12. Peng, J., Elias, J.E., Thoreen, C.C., Licklider, L.J. and Gygi, S.P. (2003) Evaluation of multidimensional chromatography coupled with tandem mass spectrometry (LC/LC-MS/MS) for large-scale protein analysis: the yeast proteome. *J. Proteome Res.* 2, 43–50.

13. Resing, K.A., Meyer-Arendt, K., Mendoza, A.M., Aveline-Wolf, L.D., Jonscher, K.R.; Pierce, K.G., Old, W.M., Cheung, H.T., Russell, S., Wattawa, J.L., Goehle, G.R., Knight, R.D. and Ahn, N.G. (2004) Improving reproducibility and sensitivity in identifying human proteins by shotgun proteomics. *Anal. Chem.* 76, 3556–3568.

14. Link, A.J., Eng, J., Schieltz, D.M., Carmack, E., Mize, G.J., Morris, D.R., Garvik, B.M. and Yates, J.R., III (1999) Direct analysis of protein complexes using mass spectrometry. *Nat. Biotechnol.* 17, 676–682.

15. Cargile, B.J., Talley, D.L. and Stephenson, J.L., Jr. (2004) Immobilized pH gradients as a first dimension in shotgun proteomics and analysis of the accuracy of pI predictability of peptides. *Electrophoresis* 25, 936–945.

16. Ishihama, Y., Rappsilber, J. and Mann, M. (2006) Modular stop and go extraction tips with stacked disks for parallel and multidimensional peptide fractionation in proteomics. *J. Proteome Res.* 5, 988–994

17. Carr, S., Aebersold, R., Baldwin, M., Burlingame, A., Clauser, K. and Nesvizhskii, A. (2004) The need for guidelines in publication of peptide and protein identification data: Working Group on Publication Guidelines for Peptide and Protein Identification Data. *Mol. Cell. Proteomics* 3, 531–533.

Chapter 9

Optical Proteolytic Beacons for In Vivo Detection of Matrix Metalloproteinase Activity

J. Oliver McIntyre and Lynn M. Matrisian

Summary

The exuberant expression of proteinases by tumor cells has long been associated with the breakdown of the extracellular matrix, tumor invasion, and metastasis to distant organs. There are both epidemiological and experimental data that support a causative role for proteinases of the matrix metalloproteinase (MMP) family in tumor progression. Optical imaging techniques provide an extraordinary opportunity for noninvasive "molecular imaging" of tumor-associated proteolytic activity. The application of optical proteolytic beacons for the detection of specific proteinase activities associated with tumors has several potential purposes: (1) Detection of small, early-stage tumors with increased sensitivity due to the catalytic nature of proteolytic activity, (2) diagnosis and prognosis to distinguished tumors that require particularly aggressive therapy or those that will not benefit from therapy, (3) identification of tumors appropriate for specific antiproteinase therapeutics and optimization of drug and dose based on determination of target modulation, and (4) as an indicator of efficacy of proteolytically activated prodrugs. This chapter describes the synthesis, characterization, and application of reagents that use visible and near infrared fluorescence resonance energy transfer (FRET) fluorophore pairs to detect and measure MMP-referable proteolytic activity in tumors in mouse models of cancer.

Key words: FRET, Dendrimer, Optical imaging, Proteolytic beacon, MMP.

1. Introduction

Fluorescence or Förster Resonance Energy Transfer (FRET) is a valuable tool in the application of optical imaging to biological problems. The process is characterized by the transfer of electronic excitation energy of a donor chromophore to an acceptor molecule brought in close proximity via a coupling mechanism between the donor and acceptor pair. The efficiency of the

Toni M. Antalis and Thomas H. Bugge (eds.), *Methods in Molecular Biology, Proteases and Cancer, Vol. 539*
© Humana Press, a part of Springer Science+Business Media, LLC 2009
DOI: 10.1007/978-1-60327-003-8_9

transfer process depends on the distance between fluorophores. The advent of optical devices, especially the CCD camera and computer-based imaging technology, has afforded tools to detect the dequenching photons from this FRET mechanism, and this process underlies the emerging molecular optical imaging approaches that enable detection of proteinase activity.

There are several examples of agents that employ FRET principles for the detection of proteolytic activity in living animals and tissues. Weissleder and colleagues attached fluorophores to a linear copolymer by a peptide containing a proteolytic cleavage site as a means to measure protease activity and subsequent inhibition in tumor-bearing mice *(1)*. Proximity of the fluorophores quenches the fluorescent signal, which is released upon cleavage of the peptide linker. Such a probe was used to detect MMP-2 activity in HT1080 human fibrosarcoma xenografts, and the signal was inhibited by treatment with a synthetic MMP inhibitor *(2)*. Tsien and colleagues described activatable cell-penetrating peptides consisting of a polyarginine membrane-translocating motif linked via an MMP-cleavable peptide to an appropriate masking polyanionic domain (a cleavable peptide hairpin) to deliver fluorescent labels within tumor cells both in vitro and in vivo after cleavage by tumor-associated proteinases *(3)*. The strategy we employed was to use a PAMAM dendrimer backbone, reference fluorophores attached directly to the dendrimer, and sensor fluorophores attached via a selective peptide linkage *(4)* (**Fig. 1**). The multivalency of the dendrimer allows for adjustment in the relative amounts of sensor and reference fluorophores, provides a vehicle that is maintained in the circulation for greater than 30 min, and provides the opportunity to link additional agents that can alter the half-life of the reagent in circulation or provide additional functionalities. The use of two different fluorophores as the FRET pair provides the opportunity to determine the ratio of cleaved product to substrate by determining the sensor:reference ratio, thereby taking issues of substrate penetration into tumors into account.

The methodology presented is for a well-characterized reagent that utilizes the visible range tetramethylrhodamine (TMR, excitation 544 nm, emission 572 nm) as the reference fluorophore, and fluorescein (excitation 494 nm, emission 519 nm) as the sensor fluorophore attached to a MMP7-selective peptide *(4)*. However, this method is adaptable for peptides that have different selectivities, and thus to monitor different enzymes or general proteolytic activity depending on the choice of peptide. In addition, the technology can be modified to utilize fluorophores in the deep red and near infrared region of the spectrum (650–850 nm), which allows enhanced visualization in biological tissues in that it minimizes interference from hemoglobin and water. **Table 1** gives the sequence of several peptides and FRET pairs that have been successfully employed in these probes.

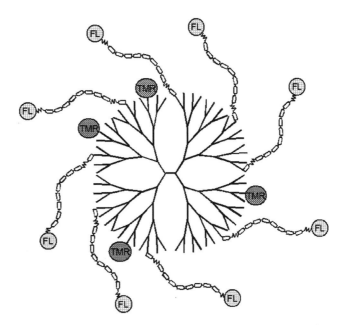

Fig. 1. Diagrammatic structure of PB-MXVIS constructed on a PAMAM generation-4 dendrimer (ethylenediamine core, diameter ~4.5 nm) with fluorescein (FL)-labeled peptide proteinase sensors, FL-MX, and tetramethylrhodamine (TMR, internal reference) linked to surface amines. For PB-M7VIS, the FL-MX peptide is FL(Ahx)RPLA*LWRS(Ahx)C-.

Table 1
Peptides and FRET pairs used in PBs

Peptide sequence	Specificity (ref.)
RPLA*LWRS	MMP7 (5)
AVRW*LLTA	MMP9 (6)
RPLG*LARE	MMP (7)
FRET sensor	**FRET reference**
Fluorescein	Tetramethylrhodamine
Cy5.5	Alexafluor750
Alexafluor700	Alexafluor750

2. Materials

2.1. Synthesis of PBs

All chemicals and biochemicals are reagent-grade and solutions are prepared in deionized filtered water (Milli-Q, Millipore Corp., www.millipore.com, Billerica, MA, USA) unless noted otherwise.

1. Generation 4 Starburst® PAMAM ethylenediamine core dendrimer is obtained as a 10% (w/w) solution in MeOH from Sigma-Aldrich (St. Louis, MO, USA)

2. *N*-succinimidyl iodoacetate (SIA) (mol. wt: 283) (Pierce Chemical, piercenet.com, Rockford, IL, USA) is dissolved to 10 mg/ml (~35 mM) in MeOH/DMF (1:1).

3. Fluorescein(FL)-(Ahx)-(octapeptide)-(Ahx)-C (FL-MX), e.g., FL-(Ahx)-RPLALWRS-(Ahx)-C (FL-M7), where Ahx is aminohexanoic acid, is an HPLC-purified peptide that includes two Ahx linkers, either from Open Biosystems (www.openbiosystems.com, Huntsville, AL, USA) or from GenScript Corp. (www.genscript.com, Piscataway, NJ, USA) and is dissolved to 8 mg/ml (~5 mM) in MeOH.

4. Tetramethylrhodamine-5-(and-6)-isothiocyanate (TRITC, mixed isomers) from Molecular Probes, Invitrogen (probes.invitrogen.com, Carlsbad, CA, USA) is prepared as a fresh solution at 5 mg/ml (~11 mM) in DMSO.

5. The *N*-hydroxysuccinimidyl (NHS) ester derivatives of the near infrared fluorophores, Alexafluor700, Alexafluor750, and Cy5.5, obtained either from Molecular Probes, Invitrogen (probes.invitrogen.com, Carlsbad, CA, USA) or from GE Healthcare (London, United Kingdom) are prepared as 7 mM solutions in DMF.

6. 0.5 M Na_2CO_3, pH adjusted to 9.0 (dilute 1:10 to measure) using HCl.

7. 0.2 M cysteine in methanol.

8. 0.2 M ethylenediaminetetraacetic acid (EDTA), pH adjusted to 8.0 with NaOH, and also diluted to 1.0 or 0.1 mM for use, as specified.

9. 0.5 M Hepes-NaOH, pH 7.0 (dilute 1:10 to measure) and diluted to 5 mM for use.

10. Glass ampoules such as 1.8-ml "Wheaton" ABC amber vial™, with Teflon/silicon lined cap (VWR International, vwr.com, Westchester, PA or Thermo Fisher Scientific Rockwood/ National Scientific, nationalscientific.com, Rockwood, TN).

2.2. Analysis of PBs

1. TMR–glycine: Add 5 µl of 5 mg/ml TRITC to 0.1 ml of 20 mM glycine in an aqueous solution.

2.3. Testing PBs In Vitro

1. Phenylmethylsulfonylfluoride (PMSF, FWt 174) is dissolved at 0.2 M in 100% ethanol and stored at 4°C. *Note*: This reagent is highly toxic.

2. 4× Tricine assay buffer stock: 0.2 M tricine [*N*-tris(hydroxymethyl)methylglycine] (>98% from Sigma-Aldrich, St. Louis, MO, USA), 0.8 M NaCl, 40 mM $CaCl_2$, 0.2 mM $ZnSO_4$, 0.02% (w/v) Brij35, adjusted to pH 7.4 with NaOH. For convenience, some of the components of this buffer are added from aqueous stock solutions, e.g., 1.0 M $CaCl_2$,

10 mM ZnSO$_4$, and 1% (w/v) Brij35 (diluted from 30% w/v solution obtained from Sigma-Aldrich). This 4× assay buffer stock solution is usually autoclaved and can be stored at ambient temperature for a number of months. Add 1 mM PMSF to the 4× tricine assay buffer before use, e.g., 50 µl of 0.2 M PMSF to 10 ml 4× buffer.

3 . Various active MMPs, abbreviated generically as MMPX, e.g., MMP2, MMP3, MMP7, and MMP9, as obtained from the supplier (usually Calbiochem, San Jose, CA, USA) are stored frozen at –80°C in appropriate aliquots (e.g., 2 or 3 µl) till required for use.

2.4. Quantitative Fluorescence Imaging of PBs

1. Matrigel™ for preparing phantoms of PB-MXVIS in the in vivo setting is from BD Biosciences (bdbiosciences.com, San Jose, CA).

2.5. In Vivo Imaging of Xenograft Tumors

1. SW480 human colon cancer cells are obtained from the ATCC (www.atcc.org, Manassa, VA, USA).

2. Preparation of PB-MXVIS for administration via i.v. injection: The stock PB-MXVIS (50–100 nmol/ml) is prepared for injection by dilution into sterile 0.9% sodium chloride solution to 1.6 nmol/150 µl kept isotonic by addition of sterile 10× PBS, as required. For example, for five animals, a total of 900 µl is prepared to provide 750 µl for injection and allowing for dead-volume losses in syringes, i.e., 120 µl of an 80 µM PB-MXVIS stock solution is added to 767 µl 0.9% saline premixed with 13 µl of sterile 10× PBS. Stock reagents that show any propensity to precipitate are routinely filtered after dilution using 13-mm 0.2-µm filters (Product no. 4602, Supor Acrodisc, Gelman Sciences).

3. Methods

3.1. Synthesis of PBs

The PBs on PAMAM dendrimer scaffolds are prepared by two sequential reactions, first linking multiple copies of sensor-labeled peptide to the reactive terminal amines of PAMAM dendrimer, followed by reaction of the PAMAM dendrimer core with reference fluorophore (*see* **Note 1**). For the first reaction, PAMAM is first activated with the bifunctional reagent, succinimidyl-iodoacetate (SIA).

1. For routine synthesis, 2.5 mg (~176 nmol) of PAMAM generation 4, i.e., 32 µl of a 10% (w/v) methanolic solution as obtained from the manufacturer, is placed in an amber glass ampoule (e.g., 1.8-ml Wheaton ABC amber vial™, with

Teflon/silicon lined cap) and allowed to react for 30 min at ambient temperature with 48-µl SIA (~3.4 µmol). The SIA/PAMAM ratio (~19) is selected to activate ~30% of the terminal primary amines calculated in PAMAM-G4 (64 surface amines/dendrimer).

2. 176 µl of FL-MX (8 mg/ml in MeOH) is added to the SIA-activated PAMAM solution to give a peptide/PAMAM ratio of 5 (total volume, 256 µl) (*see* **Note 2**). Minimize exposure of FL-MX to light, cover vial with foil, place upright on a rocking platform, and gently rock overnight at ambient temperature.

3. Remove unreacted FL-MX peptide by diafiltration after dilution with at least 8 volumes of aqueous 1 mM EDTA and ethanol to 10% with size separation using either Centriprep™ or Microcon™ centrifugal filter devices with YM-3 membranes (Amicon, Millipore Corp.) per the manufacturers' protocols and at ~8°C (*see* **Note 3**). The product (FL-MX)$_m$-PAMAM is then concentrated to ~0.5 ml after at least two rounds of diafiltration following dilution (greater than tenfold) with aqueous 1 mM EDTA. Aliquots of the original diluted reaction mixture, the effluent, washes, and retentate (product) are saved for analyses. The volume of the product, collected in a microfuge tube, is usually measured by weighing.

4. For labeling the PAMAM scaffold with TMR, (FL-MX)$_m$-PAMAM (~160 nmol) in aqueous 1 mM EDTA is made 50 mM in Na_2CO_3 (pH 9) by addition of 1/9 volume of 0.5 M Na_2CO_3 (pH 9.0), and TRITC (87 µl, ~960 nmol in DMSO) is added. After gentle rocking overnight under argon, in the dark at ambient temperature, 48 µl of 0.2 M methanolic cysteine (a tenfold excess with respect to TRITC) is added.

5. After 2 h at ambient temperature with cysteine, the reaction mixture is diluted with 8 volumes of 1 mM EDTA and 1 volume of ethanol (to 10%) and the product, (FL-MX)$_m$-PAMAM-(TMR)$_n$, is separated from unincorporated TMR by diafiltration, as described earlier, followed by at least two washes with aqueous 1 mM EDTA, a wash with dH_2O (to reduce EDTA to ~0.1 mM), and a final wash with 0.1 mM EDTA prior to concentration to about 1.0–2.0 ml for storage under argon at 4°C. For long-term storage, the beacon product, referred to as PB-MXVIS (**Fig. 1**), is adjusted to 20% ethanol (*see* **Note 4**). In addition to the product, an aliquot of the diluted reaction mixture and of the effluents from diafiltration is retained for analyses. The TMR serves both to quench the fluorescence of FL and as the internal reference.

6. Incorporating different peptides, such as those listed in **Table 1**, as the cleavable substrate in the PB, yields beacons with different substrate specificities, identified generically as PB-MXVIS.

3.2. Analysis of PBs

1. Incorporation of FL-MX into $(FL\text{-}MX)_m$-PAMAM and TMR into $(FL\text{-}MX)_m$-PAMAM-$(TMR)_n$ is calculated from the amplitude of the absorbance spectra for FL (at 497 nm) and TMR (at 554 nm), respectively. For each reaction step, the absorption spectra of the reaction mixture (after dilution into 1 mM EDTA for diafiltration), effluent, diafiltration washes, and final product (usually diluted 100 or 200-fold in 1 mM EDTA) are measured and used to calculate the incorporation of each component (FL-MX and TMR, each usually >80%) into the PAMAM dendrimer.

2. The recovery of PAMAM is measured by ninhydrin reaction by the method of Moore and Stein as described in detail elsewhere (4) and is routinely found to be ~90% in each step giving a final yield of ~80% of the starting material, i.e., ~140 nmol $(FL\text{-}MX)_m$-PAMAM-$(TMR)_n$.

3. Fluorescence excitation and emission spectra of both the $(FL\text{-}MX)_m$-PAMAM intermediate and the final $(FL\text{-}MX)_m$-PAMAM-$(TMR)_n$ product are recorded after dilution (usually 500-fold) to ~0.2 μM or to an OD < 0.1/cm (at both 497 and 554 nm) using either dH_2O or 5 mM Hepes-NaOH buffer (pH 7.0). Although accurate measurement of quantum yield and spectral corrections have not been implemented, the amplitude of the fluorescence spectrum of FL in $(FL\text{-}M7)_{4.5}$-PAMAM is ~40% of that for the same concentration of FL or FL-M7 in aqueous solution due to fluorescence quenching by homotransfer in the $(FL\text{-}M7)_{4.5}$-PAMAM dendrimer. In $(FL\text{-}M7)_{4.5}$-PAMAM-$(TMR)_{4.5}$ the FL amplitude is further attenuated (to ~6%) by FRET to TMR, and the TMR fluorescence of the beacon was ~70% of that for the same concentration of TMR–glycine.

3.3. Testing PBs In Vitro

For testing proteolytic cleavage of PB-MXVIS by various proteinases, the reagent is diluted, usually to ~0.2 μM, into buffer, dispensed in triplicate into a 96-well plate, and fluorescence of both the FL sensor and TMR reference is read as a function of time following addition of proteinases. Experimental details are as follows:

1. Prepare a "Master Mix" working solution of PB-MXVIS in tricine buffer; an aliquot of the PMSF-treated 4× tricine assay buffer is diluted with an appropriate volume of PMSF-treated H_2O, and PB-MXVIS is added to ~0.2 μM. For each 1 ml of Master Mix, mix together 500 μl of PMSF-treated 4× tricine

buffer plus 2 µl of 0.1 mM PB-MXVIS (final concentration of 0.1 µM in assay) and 498 µl of PMSF-treated H$_2$O. The volume of working solution required is dictated by the number of proteinases being tested; for assaying activity with a single proteinase, a minimum volume of 0.45-ml Master Mix is required, sufficient for nine wells (triplicate assays of three conditions – enzyme, enzyme plus either EDTA or inhibitor, and no enzyme).

2. To set up the assay plate, 50-µl aliquots of Master Mix are distributed in the required number of wells of the 96-well plate; PMSF-treated dH$_2$O is added to each well to give a total assay volume of 100 µl (e.g., 47 µl of dH$_2$O for the plus enzyme wells) and 15 µl 0.2 M EDTA or appropriate volume of inhibitor (e.g., 10 µl of 0.1 mM aqueous GM6001) (*see* **Note 5**). Before addition of enzyme, the fluorescence of both FL and TMR is read on a plate-reading fluorometer with appropriate excitation and emission filter settings.

3. Shortly before use, an aliquot of MMP stock solution is removed from the freezer, thawed, and diluted with PMSF-treated d H$_2$O to prepare a working solution, e.g., 2 ng/µl (~0.1 µM) MMP-7, 7 ng/µl (~0.11 µM) MMP-2, 5 ng/µl (~0.12 µM) MMP-3, or 7 ng/µl (0.11 µM) MMP-9. Working solutions of other proteinases, e.g., trypsin at 0.1 µg/µl, are prepared either fresh or by dilution from a stock (e.g., 1 µg/µl) stored in the freezer. Multiple freezing/thawing of stock proteinase solutions is to be avoided. Working proteinase solutions are kept on ice and are usually discarded after use, though the MMP7 working solution can be frozen in 50-µl aliquots for subsequent use without much loss in activity (*see* **Note 6**).

4. After the plate is equilibrated at the assay temperature (as programmed in the fluorometer), the plate is removed and aliquots of working solution proteinases are added to each well, as required, e.g., 3–5 µl of MMP-7 (2 ng/µl); fluorescence is then read after mixing that can be accomplished by briefly shaking the plate in the fluorometer.

5. After proteinases are added, fluorescence in both the sensor (FL) and reference (TMR) channels is read initially and then FL fluorescence is recorded at intervals, e.g., every 3 min for 10 min, every 10 min to 30 min, and then every 15 min up to ~2 h, recording the TMR channel about once or twice per hour (*see* **Note 7**). Between 1 and 2 h, the plate is removed from the fluorometer, sealed with parafilm, covered with foil, and placed in a 37°C incubator to allow the reaction to go to completion with the plate being reread at 6 h and again after overnight at 37°C.

6. The enzymatic activity is calculated from the rate of increase in FL fluorescence by plotting fluorescence versus time (**Fig. 2**) and measuring the initial or maximal slope (delta fluorescence/time, ΔF/min) that can then be converted to an equivalent rate of cleavage of the substrate from the measured maximal change in fluorescence (ΔF_{max}, taken to represent complete cleavage of the substrate). For example, from **Fig. 2**, the initial slope obtained with 0.006-μg MMP7 is 2.46 ΔF/min with a maximal change in fluorescence, ΔF_{max}, at the endpoint (not shown) of 85 for an assay using PB-M7VIS (0.1 nmol/ml) in a total volume of 80 μl, i.e., 8 pmol PB-M7VIS/assay. The calculated specific activity (SA) is given by the equation:

$$SA = [2.46\ (\Delta F/\text{min})/0.006\ (\mu gMMP7)]/$$
$$[85(\Delta F_{max})/8(\text{pmol PB-M7VIS})]$$
$$= 38.6\ \text{pmol PB-M7VIS/min/}\mu gMMP7.$$

3.4. Quantitative Fluorescence Imaging of PBs

Quantitative fluorescence imaging is achieved by digital imaging with either a full-frame, black and white CCD camera (MicroMax 1317-K/1, Princeton Instruments, www.piacton.com, Trenton, NJ, USA) coupled to a fluorescence microscope or with a whole animal fluorescence imaging system such as the IVIS 200

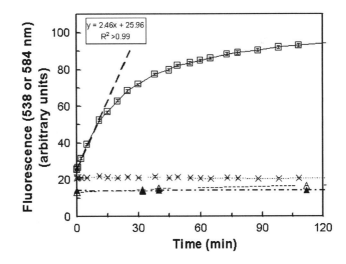

Fig. 2. In vitro assay of MMP-7 activity with PB-M7VIS as fluorogenic substrate. PB-M7VIS (0.1 μM) was incubated (37°C) with 0.006-μg MMP-7, measuring the fluorescence (microtiter plate fluorometer) of both FL (ex, 485 nm; em, 538 nm) (*open squares, crosses*) and TMR (ex, 530 nm; em, 584 nm) (*triangles*) in the absence (*open symbols*) or presence (*cross and filled triangle*) of EDTA (mean ± SEM, n = 3). The MMP-7 activity is calculated from the initial slope (*dashed line*) of the increase in FL fluorescence in the absence of EDTA and the maximal increase in FL fluorescence (usually measured at >5 h or overnight).

(Xenogen Corp., www.xenogeny.com, Hopkinton, MA, USA). Each of these imaging systems have appropriate sensitivity for quantitative detection of fluors at nanomolar concentrations with a dynamic range of more than two orders of magnitude suitable for quantitative fluorescence imaging. Before in vivo quantitative fluorescence imaging studies can be initiated, the imaging system is calibrated both in vitro and in vivo with appropriate fluorescence standards and phantoms, as described later for the microscope and IVIS imaging systems.

3.4.1. Microscope System

Quantitative fluorescence imaging is achieved by taking digital pictures with a full-frame, black and white CCD camera (MicroMax 1317-K/1, Princeton Instruments) coupled to a fluorescence microscope with a variety of Plan-Neofluar objective lenses (Axiophot, Carl Zeiss, Inc) including a 2.5× objective. Meta-Morph imaging software (Universal Imaging Corp., Downington, PA) is used to control the CCD camera and image acquisition as well as for data analyses. Data acquisition parameters and camera exposure conditions are adjusted to give fluorescence signal linear with exposure time and with largest dynamic range.

3.4.2. Calibration of Microscope Imaging

1. The fluorescence imaging system is first calibrated using a set of beacon samples prepared with the same two-color PB-MXVIS, e.g., PB-M7VIS, $(FL-M7)_m$-PAMAM-$(TMR)_n$, as used in vivo as well as controls prepared either with PB-MXVIS previously treated with MMPX or with each of the individual fluors, e.g., $(FL-MX)_m$-PAMAM and PAMAM-$(TMR)_n$. For the microscope imaging system, the reference samples are each prepared as serial dilutions, e.g., 1 µM to 10 nM, and then loaded in capillary glass tubes (0.75-mm borosilicate glass, World Precision Instruments, Sarasota, FL) sealed with hematocrit putty (Critoseal®, Krackeler Scientific, Inc., Albany, NY). After initial screening, capillaries with at least two of the concentrations of each of the compounds are mounted together in parallel on the microscope stage for calibration of the fluorescence of the two colors and assessing discrimination between the optical channels. In addition, the dynamic range of the fluorescence microscope for imaging PB-MXVIS cleavage by MMPX can be calibrated with a set of PB-MXVIS samples consisting of mixtures of uncleaved and MMPX-cleaved reagent, prepared at 0.2 nM in tricine buffer plus 20 mM EDTA and loaded into capillaries.

2. The calibration samples are oriented and focused under low-intensity white light prior to fluorescence excitation. Light is collected through one of the two low magnification objectives, depending on whether the measurement was for a calibration sample (usually 10×/0.50 Plan-Neofluar) or for in vivo

imaging of subcutaneous phantoms or tumors in mice (usually 2.5×/0.075 Plan-Neofluar). For each calibration sample of PB-MXVIS as well as control capillaries, triplicate images are acquired with at least two exposure times (usually 1, 3, 10, or 30 s) (*see* **Note 8**). The FL (fluorescein) fluorescence (green channel, 500–530-nm emission) was discriminated using a green band-pass filter set (#41025, Chroma Technology Corp. Brattleboro, VT), and the TMR (tetramethylrhodamine) fluorescence (red channel, >570-nm emission) was discriminated using a red long-pass filter set (#41032, Chroma Technology Corp.). The green channel selectively detects FL fluorescence from PB-MXVIS with only background signal from PAMAM-$(TMR)_n$, though only part of the total FL emission band can be acquired. Thus, the microscope imaging system gives different FL/TMR ratios compared with those calculated from corrected spectral amplitudes measured in a fluorometer because the long-pass filter in the red channel detects a larger fraction of the total TMR fluorescence compared with the relatively narrow FL fluorescence band collected through the green channel. This condition limits the dynamic range compared with the fluorometer experiment, but it allows effective signal discrimination in the presence of both dyes. The difference in sensitivity of the two channels can be, in part, compensated by increasing the green channel exposure with respect to the red channel, e.g., 10 and 1 s, respectively.

3. The fluorescence intensity for each of the standards and in both FL and TMR channels is calculated as the average counts/pixel after subtraction of background signals from control capillaries containing only tricine assay buffer plus 20 mM EDTA. Measurements from triplicate images of each sample are expressed as the mean (±SEM). Provided signal remains below saturation, the response should be linear with exposure time and with the concentration of both FL and TMR (*see* **Note 8**).

3.4.3. Whole Animal Imaging Systems

A number of commercial optical imaging systems have been developed for fluorescence imaging of small animals including steady-state fluorescence and bioluminescence imaging systems, e.g., the IVIS200 (Xenogen, www.xenogeny.com, Alameda, CA, USA), the Maestro spectral imaging system (CRI Inc., www.cri-inc.com, Woburn, MA, USA), the Kodak FX Pro in vivo imaging system (Carestream Molecular Imaging, www.Kodak.com, New Haven, CT, USA), and the OV100 Olympus (Leeds Precision Instruments, www.leedsmicro.com, Minneapolis, MN, USA) and time-resolved fluorescence lifetime imaging systems such as the GE Healthcare eXplore Optix (GE Healthcare, www.gehealthcare.com). The methods described later for calibrating the IVIS200 optical imaging system should, with appropriate modification depending on instrument specifications, be useful for calibrating various whole animal imaging systems. The IVIS200 is equipped with a cryogenically cooled CCD camera that was designed primarily for the sensitive quantitative detection of bioluminescence but that has been adapted for fluorescence imaging. The intrinsic sensitivity of the camera facilitates quantification of fluorescence images collected with either relatively narrow band-pass emission filters or at low fluor concentrations in vivo. For PB-MXVIS imaging in the IVIS200, the sensor (fluorescein) and reference (tetramethylrhodamine) fluorescence signals are imaged with the GFP (excitation, 445–490 nm; emission, 515–575 nm) and DsRed (excitation, 500–555 nm; emission, 575–650 nm) channels, respectively, though custom filters could also be installed. With the standard GFP and DsRed filters, fluorescence images are usually obtained with exposure times in the range from 1 to 10 s. Longer exposure times, e.g., up to 60 s, may be required for low concentrations of fluors.

3.4.4. IVIS200 Calibration

The response of the IVIS200 to the FL and TMR fluors used in PB-MXVIS is calibrated using the same strategy as for the microscope imaging system. Calibration samples, prepared as serial dilutions from both the two-color PB-MXVIS, e.g., (FL-MX)$_m$-PAMAM-(TMR)$_n$, reagent (with and without MMPX treatment) and from reference compounds containing only one of each of the individual fluors, i.e., (FL-MX)$_m$-PAMAM and PAMAM-(TMR)$_n$, are pipetted (100-µl aliquots) as an array in a black 96-well plate (#7605 Microfluor, www.thermo.com, Thermolabsystems, Franklin, MA). Fluorescence images in both sensor and reference channels are recorded with varying exposure times and for different fields of view (C & D). For quantification using the LivingImage® software provided with the IVIS, regions of interest (ROIs) are usually placed in the center of each well so as to minimize edge-of-well imaging artifacts and fluorescence, measured as photons/s/steradian, in each channel plotted versus concentration of reference beacon. The fluorescence of the individual color compounds, (FL-MX)$_m$-PAMAM and PAMAM-(TMR)$_n$, as measured in both sensor and reference channels is used to determine the efficiency of color discrimination of the two channels. With FL and TMR in the standard GFP and DsRed channels, there is ~10% overlap of TMR in the FL channel and <10% FL signal detected in the DsRed channel. With a pair of near-IR fluors such as Alexafluor680 (or Cy5.5) versus Alexafluor750, the standard Cy5.5 and ICG filters provided on the IVIS200 gave a similar discrimination between fluors with ~10% overlap of fluorescence detected in the two channels. Although the fluorescence of adjacent duplicate wells is acceptably reproducible, there is some degree of curvature in the fluorescence field, i.e., the measured signal is dependent on position within the field-of-view. For calibrating the closest field-of-view available for fluorescence (field B), an alternative sample geometry is required as the illuminating light source casts shadows within many of the wells of the 96-well plate format.

3.4.5. Calibration for In Vivo Imaging

1. For calibration of the imaging system response to PB-MXVIS in the in vivo setting, both the uncleaved and MMPX-treated reagents are prepared as phantoms in Matrigel® (from BD Biosciences, bdbiosciences.com, San Jose, CA) injected subcutaneously in anesthetized athymic nude mice (Harlan, Harlan. com, Indianapolis, IN, USA) (*see* **Note 9**). For protease treatment, PB-MXVIS stock solution is diluted to ~5 µM in tricine buffer (see earlier) and incubated overnight at 37°C with MMP7 (0.5 ng/µl). Aliquots of both uncleaved and MMPX-cleaved PB-MXVIS, prepared separately or as mixtures, e.g., 0, 25, 50, 75, and 100% cleaved, are then cooled on ice and diluted with nine volumes of Matrigel®, also stored on ice. For injection of duplicate 100-µl phantoms, usually 250-µl samples are prepared, drawn up into a previously chilled sterile syringe (e.g., a tuberculin syringe, 1.0 ml, 26G, 3/8 in., No. 309625, Becton-Dickinson), and immediately injected subcutaneously at points along the dorsal flank of the anesthetized mouse, injecting slowly to allow the Matrigel® to gel and form a localized phantom of 3–5-mm diameter containing PB-MXVIS (*see* **Note 10**).

2. About 10–15 min prior to imaging, the skin of the mouse over the subcutaneous phantoms is treated with glycerol, applied with a gauze pad. The glycerol treatment significantly improves contrast as well as the dynamic range for the fluorescent phantoms by ~20–30%, consistent with a reduction in excitation and emission scattering. The in vivo phantoms can be imaged with either the microscope system, equipped with wide-field objective or with a whole animal imaging system such as the IVIS 200. For microscope imaging, the anesthetized animal is placed on the stage and each of the phantoms is imaged first with white light, and then with fluorescence in both the FL and TMR channels. For the white-light images on the microscope stage, the surface of the skin is illuminated using a fiber-optic illuminator, e.g., Fiber-Lite High Intensity Illuminator, Series 180 (Dolan-Jenner Industries, Inc., http://www.dolan-jenner.com/, Boxborough, MA, USA). A set of three images is obtained for each field-of-view and three or more image sets are collected for each phantom, usually with at least two exposure times for each of the fluorescent images (FL and TMR) (*see* **Note 11**). Details of data acquisition and analysis with the IVIS are outlined under tumor imaging (later).

3. Data analysis of microscope images is carried out using Meta-Morph imaging software (Universal Imaging Corp., molec-ulardevices.com, Downington, PA). For each field-of-view, the white light and fluorescence images (FL and TMR and with different exposure times) are combined into an image "stack" for analysis as a group. Usually selected ROIs are drawn over specific areas of the image, e.g., referable to phantom and reference or to segment the phantom in parts. For each fluorescent image, a threshold is set to exclude all saturated pixels from the analysis (including spurious pixels that are essentially unresponsive, being always saturated) and the average intensity in each ROI for each of the fluo-rescent images is calculated. The FL and TMR fluorescence is determined after appropriate background subtraction, and the FL/TMR ratio is then calculated for each ROI in each image set and for each of the phantoms. The fluorescence data are analyzed only for exposure times that give ROIs with fluorescence below saturation.

3.5. In Vivo Imaging of Xenograft Tumors

For imaging tumor-associated proteinases activity, subcutane-ous xenograft tumors that express MMP-7 (SW480mat) are established on the rear flank of nude mice with control tumors (SW480neo) on the contra-lateral flank (*see* **Note 12**). After tumors >0.5-cm diameter develop, the animals are imaged under anesthesia, either with ketamine/xylaxine (145 and 14.5 mg/kg, respectively) or with isoflurane inhalation (3.5% induction and 1–2% maintenance). In the IVIS, the imaging bed is heated to maintain body temperature of the anesthetized animals; for imag-ing on the microscope system, a heat-lamp is used intermittently for this purpose.

3.5.1. Establishing Xenograft Tumors

1. SW480neo control and MMP7-expressing SW480mat colon cancer cells (1×10^6 cells) are implanted subcutaneously on the flanks of athymic nude mice (Harlan, Indianapolis, IN, USA) (*see* **Note 13**).

2. After 3–4 weeks of tumor growth to obtain tumors ~0.5–1.0 cm in diameter, mice are anesthetized using 2% iso-fluorane (3.5% induction, 1.5–2% maintenance) and imaged using either the microscope imaging system (as described earlier) or with a cryogenically cooled CCD camera, IVIS 200 Imaging System (Xenogen Corp., www.xenogeny.com, Alameda, CA, USA). For PB-MXVIS imaging, the filters of the IVIS are set to GFP (excitation, 445–490 nm; emis-sion, 515–575 nm) and DsRed (excitation, 500–555 nm; emission, 575–650 nm) channels to image the sensor (fluo-rescein) and reference (tetramethylrhodamine) fluorescence signals, respectively. Background (before beacon) reference

fluorescence images are obtained usually with a number of different exposure times, e.g., 1, 3, and 10 s. Prior to imaging, mice are routinely given a subcutaneous 0.5-ml bolus of sterile 0.9% saline to protect against dehydration during imaging and an ophthalmic jelly (Paralube vet ointment, Pharmaderm, www.pharmaderm.com, Duluth, GA, USA) is applied to the eyes (immediately postinjection of beacon) of mice to be imaged longer than 30 min, so as to protect the cornea from dehydration.

3.5.2. Administration of Proteolytic Beacon

PB-MXVIS prepared for injection at 1.6 nmol/150 µl in sterile 0.9% sodium chloride solution is administered by retro-orbital injection using an insulin syringe (0.3 ml, 28G, 1/2 in., No. 309300, Becton-Dickinson) that has minimal dead-volume losses and gives reproducible i.v. administration of beacon.

3.5.3. Fluorescence Imaging of MMPX Activity In Vivo

Fluorescence imaging is achieved using one or more of the quantitative fluorescence imaging systems described earlier, i.e., either a fluorescence microscope with a wide-field objective and CCD camera (*see* **Note 14**) or whole animal fluorescence imaging systems such as the IVIS 200 (Xenogen Corp.) (*see* **Notes 15** and **16**). In either imaging system, animals are first imaged before administration of PB and as soon after injection as is practical so as to provide images to validate the injection dose as measured by fluorescence in the appropriate reference channel, i.e., TMR in PB-MXVIS.

1. PB-MXVIS (1.6 nmol in 100 µl of sterile 0.9% saline) is injected in the retro-orbital vascular bed and animals are usually imaged as soon as practical after injection and then for either about 30 min or, for assessment of rate of clearance, up to about 60–90-min postinjection. Additional image sets of animals are recorded every hour for up to usually 4–5 h postinjection of PB-MXVIS (*see* **Note 15**). Between imaging sessions, animals are allowed to recover from anesthesia, keeping them warm and hydrated at all times (*see* **Note 17**).

2. Image data sets of tumors obtained with the microscope/CCD camera imaging system are analyzed using Meta-Morph® software as described earlier for in vivo phantoms. The IVIS imaging data sets are analyzed using Living Image® software by Xenogen in the GFP and DsRed channels that predominately measure FL (sensor) and TMR (reference) fluorescence, respectively. ROIs are created to measure the average radiance (photons/s/cm²/steradian) both pre- and postinjection of PB-MXVIS in the tumor-bearing regions as well as an appropriate control region or regions, such as the hind leg of the mouse (muscle tissue). In addition, a region

or regions of the background usually adjacent to one or more animals is also measured over time to monitor any unanticipated changes in instrument response or detection parameters. From each of the two IVIS fluorescence imaging channels, sensor (S) and reference (R) signal is measured either as signal above preinjection background and/or signal minus background postinjection (*see* **Note 18**). For both tumor and controls, the S/R ratio is calculated as a function of time after injection of the PB (*see* **Notes 12, 19, and 20**).

3.5.4. Microscopic Analysis of Tumors

The distribution of PB-MXVIS and related beacons within tumors can be assessed in frozen sections:

1. At an appropriate time after i.v. administration of the beacon (usually in the range from 1 to 4 h), mice are sacrificed by carbon dioxide asphyxiation; tumors are resected and immersed in OCT embedding compound (Tissue-Tek, www.emsdiasum.com, Hatfield, PA, USA) and quickly frozen under liquid nitrogen.

2. Frozen sections (5–10 μm) are prepared using a cryomicrotome and stored at –20°C prior to analysis. For histological analysis, OCT is removed from the samples by immersing the slides in H_2O followed by 70% EtOH. The slides are aqueously mounted in Gel/Mount aqueous mounting medium (Biomeda, www.biomeda.com, Foster City, CA, USA) containing 4′,6-diamidino-2-phenylindole (DAPI, 2 μM) for visualization of cell nuclei.

3. For quantitative fluorescence imaging, digital images are recorded with a full-frame, CCD camera (MicroMax 1317-K1; Princeton Instruments, Trenton, NJ, USA) coupled to a fluorescence microscope (see earlier) equipped with a variety (10×, 20×, and 40× oil immersion) of Plan-Neofluar objective lenses (Axiophot; Carl Zeiss, www. zeiss.com, Thornwood, NY, USA). The 40× objective lens is used under oil immersion. Imaging parameters are optimized to obtain fluorescence images linear with camera exposure time and with maximum dynamic range. White light and DAPI images are used to focus and orient the specimen field before fluorescence excitation that is usually acquired in both FL sensor and TMR reference channels (see earlier) using two or more exposure times.

4. The images are analyzed using MetaMorph software as described earlier and, for each ROI, intensity in each channel is calculated as the average counts/pixel after subtraction of background signals from control samples prepared without PB-MXVIS beacon.

4. Notes

1. Most of the synthetic steps are routinely carried out with minimal exposure to light (amber vials and/or foil wrapping) and under argon atmosphere to reduce exposure to oxygen.

2. Efficient coupling of ~8 peptides/PAMAM has also been obtained following addition of 10 equivalents of FL-MX peptide to SIA-activated PAMAM.

3. Centrifugal filter devices with YM-10 membranes may also be used.

4. In the original procedure *(4)*, both the intermediate and final products are diafiltered with 0.1 M NaCl, 5 mM Hepes-NaOH (pH 7.0), and 1 mM EDTA. While beacons, stored with or without 20% ethanol, are stable for several weeks at 4°C, reagents were also stable without saline, i.e., in 1 mM EDTA. A fraction of the reagent may precipitate over time, a process usually reversed by simple vortex mixing.

5. Although data from multiwell plate assays can be obtained with assay sample volumes as low as 50 μl/well, the intrasample and intersample variability is markedly reduced in assays using at least 80 μl/well, and 100-μl assay volumes are routinely used. The reading parameters for the plate-reading fluorometer should be tested with fluorescein, tetramethylrhodamine, and PB-MXVIS before assays are run so as to optimize: (a) filter selection to provide maximal sensitivity for FL and minimal detection of FL in the TMR channel; and (b) other data collection parameters such as integration time so as to provide rapid data acquisition within constraints of signal-to-noise of the data. A fluorometer with temperature-regulated stage is optimal, though most proteinases are also active at ambient temperature.

6. When testing cleavage of a newly synthesized beacon with a variety of proteinases, it is recommended to confirm that each of the proteinases are active by setting up a number of assays either in the same plate or a second plate with both dye-quenched(DQ)-collagen and/or DQ-gelatin, usually with these substrates at 10 μg/ml *(8)*.

7. In multiwell plate assays, collecting sets of fluorescence intensities at various points over time provides better signal-to-noise than kinetic traces of each well, where sample bleaching can occur to some extent due to prolonged exposure to the excitation light.

8. For in vitro and in vivo calibration, prolonged exposure of calibration samples, particularly those prepared in capillaries,

to the excitation light may result in partial bleaching of the fluorescence. It is recommended that, after initial adjustments have been made, in vitro calibration be performed using fresh samples in capillaries.

9. All animal experiments are carried out in accord with and after approval of the procedures and protocol by the Institutional Animal Care and Use Committeee (IACUC).

10. For initial in vivo calibration, two animals are recommended, each prepared with a maximum of eight phantoms: one mouse with phantoms prepared from four different reagents, uncleaved and cleaved PB-MXVIS (each in duplicate), and the single-color $(FL\text{-}MX)_m$-PAMAM and PAMAM-$(TMR)_n$ compounds, plus a control phantom without beacon; the second mouse is prepared with phantoms consisting of mixtures of uncleaved and cleaved PB-MXVIS.

11. The in vivo calibration with subcutaneous PB-MXVIS in Matrigel should be carried out by imaging the animals (see later) soon after implantation of the phantoms since the signal decreases over time (several hours). However, photobleaching of in vivo phantoms is generally negligible.

12. To validate in vivo targeting specificity, animal models should be designed to include a negative control such as tumor cells that have been genetically manipulated to either overexpress or knockdown the target of interest. A valuable approach to demonstrate the in vivo targeting specificity is to make use of animals made null for the target of interest so that response can be compared in positive versus null animals (preferably littermates).

13. For subcutaneous xenograft tumors, it is important to avoid placement over the kidneys or in their immediate vicinity, as a major portion of PB-MXVIS is cleared by excretion and the kidneys retain significant fluorescence for an extended period (1–2 days) after injection of PB-MXVIS.

14. In the microscope system, due to the relatively small field-of-view, several imaging sets must be recorded for each mouse, i.e., SW480neo and SW480mat tumors as well as a control region (usually adjacent to the spine, between the xenografts), and at multiple time points, usually limiting data acquisition to two or three animals that can be imaged alternately over a period of a few hours.

15. Although the IVIS is reasonably adjusted to "flat field" for bioluminescence images, fluorescence images are somewhat dependent on location within the field and it is important when recording sequential images over time that each of the animals be positioned reproducibly within the field-of-view.

16. The IVIS200 whole animal imaging system is equipped with isoflurane inhalation anesthesia for simultaneous imaging of up to five mice (field of view D). However, imaging smaller numbers of animals (three or two) using one of the smaller field-of-view settings provides increased sensitivity.

17. For most measurements recorded over time, treatment of the skin with glycerol to enhance the detection of fluorescence from the subcutaneous xenograft tumors was not implemented so to avoid introduction of an uncontrolled variable, i.e., dehydration/rehydration of the skin over the course of the study.

18. Mice can be maintained on a low fluorescence diet (TD-97184, Harlan Teklad, Madison, WI, USA) for three or more days prior to imaging to reduce background fluorescence (9).

19. Importance of timepoint for S/R analysis: Results from studies conducted with PBs in various tumor models, e.g., SW480 colon tumor xenografts *(4, 9)*, *Min* mouse adenomas *(9)*, and LLC lung tumors *(10)*, have revealed the importance of carrying out exploratory studies to measure the S/R ratios as a function of time after administration of the PB. Although the half-times for whole-body clearance of the PBs studied thus far are similar (on the order of 60–90 min), maximal S/R ratios in tumors have been found at different times after injection. With the prototype PB-M7VIS, optimal S/R ratio was at ~2-h postinjection, whereas with the second generation PB-M7NIR studied in the same SW480 tumor model system, the maximal S/R ratio was at ~4-h postinjection, a temporal difference that may in part be attributed to enhanced signal-to-noise and/or signal-to-background ratios for the NIR versus VIS versions of the PB. However, for imaging PB-M7NIR in the intestinal adenomas of the *Min* mouse, optimal S/R ratios were found at ~1-h postadministration of the PB, i.e., much sooner than in the SW480 xenograft model, a result that may be due, at least in part, to a faster wash out of cleaved sensor from the adenomas as compared with the xenografts.

20. Controls for specificity: Appropriate controls for studies with these kinds of PBs both in vitro and in vivo include the use of control PBs constructed with a scrambled peptide sequence, an uncleavable peptide of the same sequence except with d-amino acids, or with a peptide designed to target a different proteinase (*see* **Table 1**). Some validation of targeting specificity can be obtained by judicious use of pharmacological inhibitors such as the MMP inhibitors (MMPI). However, it should be noted that while the attenuation of response to a PB afforded by treatment with a pharmacological agent such

as an MMPI is indicative of biological efficacy of the drug, the attenuation of PB signal may be a result of an indirect downstream response to the agent rather than direct inhibition of the proteinase target.

Acknowledgment

This work was supported in part by RO1 CA84360.

References

1. McIntyre, J. O. and Matrisian, L. M. (2003) Molecular imaging of proteolytic activity in cancer. *J. Cell. Biochem.* 90, 1087–1097.

2. Bremer, C., Tung, C. H., and Weissleder, R. (2001) In vivo molecular target assessment of matrix metalloproteinase inhibition. *Nat. Med.* 7, 743–748.

3. Jiang, T., Olson, E. S., Nguyen, Q. T., Roy, M., Jennings, P. A., and Tsien, R. Y. (2004) Tumor imaging by means of proteolytic activation of cell-penetrating peptides. *Proc. Natl. Acad. Sci. USA* 101, 17867–17872.

4. McIntyre, J. O., Fingleton, B., Wells, K. S., Piston, D. W., Lynch, C. C., Gautam, S., et al. (2004) Development of a novel fluorogenic proteolytic beacon for in vivo detection and imaging of tumour-associated matrix metalloproteinase-7 activity. *Biochem. J.* 377, 617–628.

5. Welch, A. R., Holman, C. M., Browner, M. F., Gehring, M. R., Kan, C. C., and Vanwart, H. E. (1995) Purification of human matrilysin produced in *Escherichia coli* and characterization using a new optimized fluorogenic peptide substrate. *Arch. Biochem. Biophys.* 324, 59–64.

6. Chen, E. I., Li, W., Godzik, A., Howard, E. W., and Smith, J. W. (2003) A residue in the S2 subsite controls substrate selectivity of matrix metalloproteinase-2 and matrix metalloproteinase-9. *J. Biol. Chem.* 278, 17158–17163.

7. Kraft, P. J., Haynes-Johnson, D. E., Patel, L., Lenhart, J. A., Zivin, R. A., and Palmer, S. S. (2001) Fluorescence polarization assay and SDS-PAGE confirms matrilysin degrades fibronectin and collagen IV whereas gelatinase A degrades collagen IV but not fibronectin. *Connect. Tissue Res.* 42, 149–163.

8. Menon, R., McIntyre, J. O., Matrisian, L. M., and Fortunato, S. J. (2006) Salivary proteinase activity: a potential biomarker for preterm premature rupture of the membranes. *Am. J. Obstet. Gynecol.* 194, 1609–1615.

9. Scherer, R. L., VanSaun, M. N., McIntyre, J. O., and Matrisian, L. M. (2008) Optical imaging of matrix metalloproteinase-7 activity in vivo using a proteolytic nanobeacon. *Mol. Imaging* 7, 118–131.

10. Acuff, H. B., Carter, K. J., Fingleton, B., Gorden, D. L., and Matrisian, L. M. (2006) Matrix metalloproteinase-9 from bone marrow-derived cells contributes to survival but not growth of tumor cells in the lung microenvironment. *Cancer Res.* 66, 259–266.

Chapter 10

Dissecting the Urokinase Activation Pathway Using Urokinase-Activated Anthrax Toxin

Shihui Liu, Thomas H. Bugge, Arthur E. Frankel, and Stephen H. Leppla

Summary

Anthrax toxin is a three-part toxin secreted by *Bacillus anthracis*, consisting of protective antigen (PrAg), edema factor (EF), and lethal factor (LF). To intoxicate host mammalian cells, PrAg, the cell-binding moiety of the toxin, binds to cells and is then proteolytically activated by furin on the cell surface, resulting in the active heptameric form of PrAg. This heptamer serves as a protein-conducting channel that translocates EF and LF, the two enzymatic moieties of the toxin, into the cytosol of the cells where they exert cytotoxic effects. The anthrax toxin delivery system has been well characterized. The amino-terminal PrAg-binding domain of LF (residues 1–254, LFn) is sufficient to allow translocation of fused "passenger" polypeptides, such as the ADP-ribosylation domain of *Pseudomonas* exotoxin A, to the cytosol of the cells in a PrAg-dependent process. The protease specificity of the anthrax toxin delivery system can also be reengineered by replacing the furin cleavage target sequence of PrAg with other protease substrate sequences. PrAg-U2 is such a PrAg variant, one that is selectively activated by urokinase plasminogen activator (uPA). The uPA-dependent proteolytic activation of PrAg-U2 on the cell surface is readily detected by western blotting analysis of cell lysates in vitro, or cell or animal death in vivo. Here, we describe the use of PrAg-U2 as a molecular reporter tool to test the controversial question of what components are required for uPAR-mediated cell surface pro-uPA activation. The results demonstrate that both uPAR and plasminogen play critical roles in pro-uPA activation both in vitro and in vivo.

Key words: Anthrax toxin, Plasminogen, Protective antigen, Urokinase plasminogen activator, Urokinase plasminogen activator receptor.

1. Introduction

Anthrax toxin is a major virulence factor secreted by *Bacillus anthracis*, consisting of three polypeptides: a cellular receptor binding component – protective antigen (PrAg), and two enzymatic moieties – edema factor (EF) and lethal factor (LF) *(1)*.

Toni M. Antalis and Thomas H. Bugge (eds.), *Methods in Molecular Biology, Proteases and Cancer, Vol. 539*
© Humana Press, a part of Springer Science + Business Media, LLC 2009
DOI: 10.1007/978-1-60327-003-8_10

These three proteins are individually nontoxic. To intoxicate host mammalian cells, PrAg binds to cell surface tumor endothelial marker 8 (TEM8) or capillary morphogenesis gene 2 product (CMG2), the two widely expressed anthrax toxin receptors *(2, 3)*, and is then proteolytically activated by cell surface furin, releasing the amino-terminal 20-kDa peptide (PrAg20), thereby allowing the cell bound carboxyl-terminal 63-kDa peptide PrAg63 to form a heptamer *(1)*. Oligomerization of PrAg63 also generates EF and LF binding sites, which span the subunit–subunit interfaces on the PrAg heptamer *(4, 5)*. Thus, EF and LF can only bind to the oligomeric form, not the monomeric form of PrAg63. Under saturating conditions, one PrAg63 heptamer can bind a maximum of three molecules of LF or EF. Oligomerization of PrAg63 not only provides the binding site for LF and EF, but also triggers internalization of the toxin complex into endosomes via a lipid raft-mediated clathrin-dependent process *(6, 7)*. A decrease in the pH in endosomes causes the PrAg63 heptamer to insert in endosomal membranes to form a channel, and through this channel LF and EF translocate to the cytosol to exert their cytotoxic effects. Therefore, PrAg is the central part of anthrax toxin, serving as the delivery vehicle for binding and translocation of LF and EF into the cytosol of the cells. The combination of PrAg and LF, termed lethal toxin (LeTx), kills animals *(8, 9)* and certain cells, including some murine macrophages *(10)*, but shows no evident cytotoxicity to most other cell types. LF is a zinc-dependent metalloprotease that cleaves several mitogen-activated protein kinase kinases (MAPKKs) in their amino-terminal regions *(11, 12)*. The combination of PrAg and EF, called edema toxin, causes edema when injected subcutaneously and death when injected systemically in experimental animals *(13)*. EF is a calcium- and calmodulin-dependent adenylate cyclase, which elevates intracellular cAMP concentrations *(14)*, thereby causing diverse effects in cells including the impairment of phagocytosis *(15)*.

The amino-terminal sequence of LF (residues 1–254, LFn) has substantial sequence homology to the amino-terminal sequence of EF *(1)*. This region constitutes the PrAg hetpamer-binding domain and is sufficient to allow translocation of fused "passenger" polypeptides to the cytosol of cells in a PrAg-dependent process *(16–18)*. Thus, LFn fused to other bacterial toxin enzymatic domains such as the ADP-ribosylation domain of *Pseudomonas* exotoxin A (fusion protein 59, or in short, FP59) *(17)*, or to reporter enzymes, such as β-lactamase (LFnLac) *(19)*, have been generated. FP59 can be used as a potent antitumor agent when delivered to tumor cells using a tumor-specific PrAg *(20–22)*. LFnLac was successfully used to image cells expressing various proteases when combined with the protease-specific PrAg proteins *(19)*.

The unique requirement for PrAg proteolytic activation on the target cell surface provides a way to reengineer this protein to make its activation dependent on proteases other than furin that are present on the surface of the target cell. To this end, we have generated PrAg variants that are selectively activated by urokinase plasminogen activator (uPA) *(21)*, a serine protease that is overproduced along with its cognate receptor (uPAR) by a variety of tumor tissues and tumor cell lines. Another physiological plasminogen activator, tissue plasminogen activator (tPA), shares with uPA an extremely high degree of structural similarity and the same primary physiological substrate (plasminogen) and inhibitors (PAI-1 and PAI-2). Unlike uPA, which normally functions in tumor cells, tPA is expressed and secreted mostly by vascular endothelial cells and is primarily involved in clot dissolution. Therefore, one concern in the design of uPA-dependent PrAg proteins is to avoid cross-activation by tPA in order to minimize the potential toxicity to blood vessels. Successful discrimination of substrate sequences between uPA and tPA was made possible by the work of Madison and colleagues *(23, 24)* who used phage display to identify peptide sequences that are cleaved with high efficiency and selectivity by either uPA or tPA. Thus, using an optimized uPA substrate sequence GSGRSA to replace the furin site RKKR in PrAg yielded PrAg-U2, a PrAg variant that is efficiently and preferentially activated by uPA *(21, 22)*. In contrast, when the furin site was replaced by a tPA-preferred recognition sequence, QRGRSA, the resulting PrAg-U4 was preferentially activated by tPA *(21)*. In theory, the PrAg furin cleavage site RKKR can be changed to any other protease cleavage site to generate a PrAg cleaved by a particular protease, provided a specific substrate sequence is known. In this chapter, we will focus on PrAg-U2 protein purification and its usefulness in dissecting urokinase activation pathways both in vitro and in vivo. Using the modified anthrax toxin to image cell surface protease activities is described in chapter 7 (Hobson et al.). For other applications of the modified anthrax toxins, please refer to *(25)*.

2. Materials

2.1. Nonvirulent B. anthracis Protein Expression System

1. Expression plasmids: pYS5 is a PrAg expressing plasmid that can shuttle between *E. coli* and *B. anthracis* *(26)*, allowing molecular cloning to be done in *E. coli* and protein expression and purification in nonvirulent *B. anthracis* strain BH450. In this plasmid, the expression of PrAg is driven by the original PrAg promoter. To make PrAg-U2, the DNA sequence encoding the PrAg furin cleavage sequence in pYS5 was changed to that encoding the uPA cleavage

peptide PGSGR ↓SA (↓indicates cleavage site), resulting in an uPA-activated PrAg expressing plasmid pYS-PrAg-U2 *(21)*. To express FP59 in this system, the mature PrAg coding sequence in pYS5 was replaced with the FP59 coding sequence, resulting in pYS-FP59, which expresses FP59 with the PrAg signal peptide at the amino terminus *(27)*. All other protease-specific PrAg proteins can be efficiently made using this system.

2. Host strain for expression: BH450 is a protease and sporulation-deficient, virulence plasmid-cured *B. anthracis* strain previously designated MSLL33 *(28)*. The expression plasmids are transformed into BH450 by electroporation (the electroporation transformation protocol is available upon request).

2.2. FA Medium

FA medium is used to culture BH450 transformants for protein expression and purification.

1. Enriched FA medium (1 L): mixture of 900mL FA medium (35g Bacto tryptone, 5g Bacto yeast extract, autoclaved to sterilize) and 100 mL of 10× salts.

2. 10× salts (1 L): 60g $Na_2HPO_4 \cdot 7H_2O$, 10g KH_2PO_4, 55g NaCl, 0.4g L-tryptophan, 0.4g L-methionine, 0.05g thiamine–HCl, 0.25g uracil; adjust pH to 7.5, and filter to sterilize.

2.3. Protein Purification

1. Phenyl-Sepharose Fast Flow (low substitution) resin (GE Healthcare Life Sciences): Store in 20% ethanol at 4°C. The used resin can be recycled by sequentially washing with 10 volumes of 0.1N NaOH and large amounts of distilled water, and then stored in 20% ethanol at 4°C.

2. Q-Sepharose Fast Flow resin (GE Healthcare Life Sciences).

3. Ammonium sulfate (Sigma, St. Louis, MO): Solid ammonium sulfate is precooled at –20°C before use.

4. Phenylmethylsulfonyl fluoride (PMSF): 30 mg/mL stock solution in isopropanol; store at –20°C.

5. 500 mM stock solution of EDTA in H_2O, pH 7.4.

6. Washing buffer: 1.5 M ammonium sulfate, 10 mM Tris–HCl, 1 mM EDTA, pH 8.0.

7. Elution buffer: 0.3 M ammonium sulfate, 10 mM Tris–HCl, pH 8.0, 0.5 mM EDTA.

8. Buffer A: 10 mM Tris–HCl, pH 8.0, 1 mM EDTA, pH 8.0.

9. Buffer B: Buffer A with 0.5 M NaCl.

2.4. Reagents

1. Rabbit anti-PrAg serum #5308 (made in our laboratory) can recognize various PrAg species when used in western blotting. A 1:5,000 dilution can be used in western blotting.

2. Human pro-uPA (single-chain uPA) (no. 107), monoclonal antibody against human uPA B-chain (no. 394), PAI-1 (no. 1094),

and Glu-plasminogen (no. 410) (American Diagnostica, Inc., Greenwich, CT).

3. Tranexamic acid and MTT (3-[4,5-dimethylthiazol-2-yl]-2,5-diphenyltetrazolium bromide (Sigma).

2.5. Cell Culture and Western Blotting

1. HeLa cells and human 293 kidney cells obtained from American Type Culture Collection are grown in Dulbecco's modified Eagle's medium (DMEM) (Invitrogen) supplemented with 10% fetal bovine serum (Invitrogen), 0.45% glucose, 2 mM glutamine, and 50 μg/mL gentamicin.

2. Solution of trypsin (0.25%) and EDTA (1 mM) (Invitrogen).

3. Hanks' balanced salt solution (HBSS) (Biofluids, Rockville, MD).

4. Modified Radioimmunoprecipitation buffer (RIPA): 50 mM Tris–HCl, pH 7.4, 1% Nonidet P-40, 0.25% sodium deoxycholate, 150 mM NaCl, 1 mM EDTA. May be stored up to 1 year if filter sterilized and stored at 4°C in the dark. RIPA buffer may be supplemented with complete protease inhibitor cocktail tablet (see later) immediately before use.

5. Complete protease inhibitor cocktail tablets, from Roche Diagnostics (Mannheim, Germany).

6. 4–20% gradient Tris–glycine gels (Novex) from Invitrogen.

7. 6× SDS sample buffer: 0.35 M Tris–HCl, pH 6.8, 10% SDS, 36% glycerol, 0.6 M dithiothreitol, 0.01% bromphenol blue. Store at –20°C in aliquots.

2.6. Mice

1. uPA knockout mice as described *(29)*.

2. uPAR knockout mice as described *(30)*.

3. Plg (plasminogen) knockout mice as described *(31)*.

4. PAI-1 knockout mice as described *(32)*.

3. Methods

uPA and uPAR are overexpressed by virtually all human tumors and can be considered as a hallmark of malignant conversion *(33, 34)*. uPA and uPAR are expressed at very low levels in normal tissues, but their expression is rapidly induced in response to tissue injury, thereby providing extracellular proteolysis essential for tissue repair and remodeling *(34–37)*. uPA is secreted as a single chain enzyme (pro-uPA) with very low intrinsic activity, and is converted to the active form, two-chain uPA, by plasmin *(38)*. Two-chain uPA, in turn, is a potent activator of plasminogen (Plg), by cleaving the R^{560}-V^{561} site in plasminogen, giving rise to active plasmin. The pivotal role of uPAR in uPA-mediated cell surface plasminogen activation is well defined biochemically *(34)*, but the function of uPAR in vivo was recently challenged by the milder phenotype of uPAR$^{-/-}$ mice compared with uPA$^{-/-}$ mice *(30, 39, 40)*. By taking advantage of the facts that uPA-dependent activation of PrAg-U2 is readily detected by immunoblotting in vitro, or as a cause of cell death in vivo, here we describe the use of PrAg-U2 as a molecular tool to test the established paradigms regarding uPAR-mediated cell surface uPA activation both in vitro and in vivo. The results demonstrate that both uPAR and plasminogen play critical roles in pro-uPA activation both in vitro and in vivo.

3.1. Expression and Purification of Prag-U2

1. The BH450 bacteria transformed with the PrAg-U2 expressing plasmid pYS-PrAg-U2 are grown from an inoculum of several resuspended colonies in six 3-L flasks, each containing 500mL FA medium with 10 µg/mL kanamycin for 12–15 h at 37°C with shaking at 220 rpm (*see* **Note 1**).

2. Place flasks in ice-water bath, and add PMSF to 10 µg/mL to the cultures. The culture supernatants are then collected by centrifugation at 4,500 × *g* for 30 min at 4°C. The supernatants are sterilized by pumping through a 0.2-µm cartridge filter (Millipak 60, Millipore Corp., Bedford, MA). Add EDTA (5 mM) to further minimize protein degradation. All the following steps should be done in a cold room.

3. The proteins secreted into the culture supernatants are then precipitated on Phenyl-Sepharose Fast Flow resin in the presence of 2 M ammonium sulfate. Divide the 3-L sterile supernatant between two 3-L tissue culture roller bottles, and add precooled solid ammonium sulfate (270 g per liter supernatant) and Phenyl-Sepharose Fast Flow resin (50mL settled resin to each bottle).

4. Gently rotate the bottles until ammonium sulfate is dissolved, and then at least for 1 h more.

5. Collect resin on a porous plastic funnel (Bel-Art Plastics, 8-cm diameter). The porous filter should be wetted with ethanol and then washed with water to remove air.

6. Wash the resin on the funnel with 500mL washing buffer.

7. The proteins are then eluted using elution buffer. Add this initially in 10-mL portions, dropwise, to achieve laminar flow as if this were a chromatography column. Collect fractions of 10–20 mL until pigment (and protein) elute(s).

8. Collect approximately 300mL protein elute, place in two 200-mL centrifuge bottles (150mL protein elute in each), and precipitate proteins by adding solid ammonium sulfate to 70% saturation (30g solid ammonium sulfate per 100 mL of eluate). Rotate the bottles until ammonium sulfate is dissolved, and then leave for at least additional 1 h (it is convenient to leave this overnight).

9. Centrifuge in 200-mL bottles, 9,000 × g, 20 min. Pour off the supernatant. Dissolve the protein pellet in 20mL 10 mM Tris–HCl, 1.0 mM EDTA, pH 8.0, and dialyze the solution >5 h against 10 mM Tris–HCl, 1 mM EDTA, pH 8.0.

10. PrAg-U2 is further purified by chromatography on a Q-Sepharose FF column using an AKTA Purifier 10 FPLC system (GE Healthcare Life Sciences) or equivalent system. The column is 1.5-cm diameter, 15-cm long. Approximately 20mL dialyzed protein solution from last step is loaded on the column and washed with 300mL buffer A.

11. PrAg-U2 is then eluted from the column using 300mL 0–50% gradient buffer B, pumped at 1 mL/min, and fractions are collected.

12. Run SDS- or native-PAGE to identify the fractions containing PrAg-U2 (see **Note 2**). Pool the fractions with PrAg-U2, and dialyze against 5 mM Hepes, 0.5 mM EDTA. Filter sterilize, measure UV spectrum, calculate mg/mL as $A_{280} \times 1.09$. Freeze aliquots (see **Note 3**).

13. FP59 can be purified to one prominent band with expected size of 53 kDa from the culture supernatant using the same procedures as described earlier.

3.2. Proteolytic Activation of Pro-uPA and PrAg-U2 on Cultured Cells

1. Cells (such as uPAR-expressing HeLa cells and uPAR-non-expressing human 293 cells) are seeded in 24-well plates to allow them to grow near confluence (80–100%) the next day (see **Note 4**).

2. The cells at 80–100% confluency are washed once with HBSS, followed by incubation in 1 mL/well serum-free DMEM containing 1 µg/mL pro-uPA, 1 µg/mL Glu-plasminogen, 1 µg/mL PrAg-U2, and 2 mg/mL bovine serum albumin (BSA) at 37°C for various lengths of times (**Fig. 1**).

3. When plasminogen activator inhibitor-1 (PAI-1) is tested, cells are preincubated with PAI-1 for 30 min prior to the addition of pro-uPA, Glu-plasminogen, and PrAg-U2. When tranexamic acid is used to strip the cell surface-bound plasminogen, cells are preincubated with serum-free DMEM containing 2 mg/mL BSA, 1 mM tranexamic acid, without plasminogen, for 30 min before the addition of pro-uPA and PrAg-U2 (see **Note 5**).

4. Cell culture plates are then placed on ice, and washed five times with precooled (on ice) HBSS to remove unbound pro-uPA, PrAg-U2, and other additions (such as inhibitors), and then lysed in 100 µL/well of modified RIPA lysis buffer supplemented with complete protease inhibitor cocktail tablet on ice for 10 min (see **Note 6**).

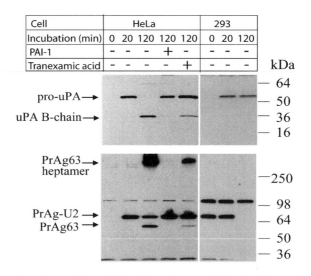

Cell	HeLa					293		
Incubation (min)	0	20	120	120	120	0	20	120
PAI-1	−	−	−	+	−	−	−	−
Tranexamic acid	−	−	−	−	+	−	−	−

Fig. 1. Binding and processing of pro-uPA and PrAg-U2 by HeLa and 293 cells. HeLa and 293 cells were cultured to confluence in 24-well plates, and preincubated with serum-free DMEM containing 2 mg/mL BSA, 1 μg/mL of Glu-plasminogen, with or without 10 μg/mL of PAI-1 for 30 min. Some cells were preincubated with serum-free DMEM containing 2 mg/mL BSA, 1 mM tranexamic acid, without plasminogen. Then 1 μg/mL each of pro-uPA and PrAg-U2 were added to the cells and incubated for the times indicated. The cells were thoroughly washed, and the cell lysates were analyzed by western blotting using a monoclonal antibody against the uPA B-chain (#394) (*upper* panel), or by using a rabbit anti-PrAg polyclonal antibody (#5308) (*lower* panel) to determine the binding and processing status of pro-uPA and PrAg-U2.

5. Mix 50μL cell lysate from each well with 10 μL 6× SDS sample buffer, heat at 95°C for 5 min, and vortex vigorously to break genomic DNA before sample loading (*see* **Note 7**).

6. 10–15 μL samples from each well along with a protein molecular weight marker are loaded onto 4–20% gradient Tris–glycine gels to run SDS-PAGE at 120 V (*see* **Note 8**). It takes approximately 2 h for the dye to reach bottom of the gel.

7. Proteins on the gel are then transblotted onto nitrocellulose membranes using any method that is successful in your laboratory.

8. Western blotting is performed to detect pro-uPA and active form of uPA B-chain using a monoclonal antibody against human uPA B-chain (no. 394, American Diagnostica, 1:1,000 dilution) and goat anti-mouse IgG (HRP conjugate, preabsorbed, Santa Cruz, 1:2,000 dilution) following the universal protocols described in either Upstate Biotechnology or Santa Cruz Biotechnology Immunoblotting protocols (*see* **Note 9**).

9. To detect the proteolytically processed products of PrAg-U2, the same set of samples run on anther gel are probed with 1:5,000 dilution of a rabbit anti-PrAg antiserum (#5308), followed by a donkey anti-rabbit IgG (HRP conjugate, preabsorbed, Santa Cruz, 1:2,000 dilution) (*see* **Note 9**).

10. The results of an example experiment were shown in **Fig. 1**. HeLa cells proteolytically activated pro-uPA on the cell surface (the appearance of uPA B-chain at 120 min in SDS gel). In contrast, the uPAR nonexpressing human 293 kidney cells bound weakly (probably unspecific binding) and could not proteolytically activate pro-uPA. The activation of pro-uPA by HeLa cells was completely blocked by PAI-1. Activation of PrAg-U2 on HeLa cell surface, determined by the production of the processed form PrAg63 and the formation of SDS-stable PrAg63 heptamer, exactly matched the activation profile of pro-uPA. In particular, when the activation of pro-uPA was blocked by PAI-1, or by the use of tranexamic acid, which inhibits the binding of plasminogen to the cell surface, PrAg-U2 activation was blocked in parallel. These results demonstrate that the activation of pro-uPA requires simultaneous binding of pro-uPA and plasminogen to cell surface.

3.3. Cytotoxicity of PrAg-U2/FP59 To uPAR-Expressing Cells

1. uPAR-expressing cells (such as HeLa cells) are seeded into 96-well plates and grown to 30–50% confluence (*see* **Note 10**).

2. The cells are washed twice with serum-free DMEM to remove residual serum. Then the cells are preincubated for 30 min with serum-free DMEM containing 100 ng/mL pro-uPA and 1 μg/mL Glu-plasminogen with or without PAI-1 (*see* **Note 11**). Various concentrations of PrAg-U2 (0–1,000 ng/mL) combined with FP59 (constant at 50 ng/mL) are added to the cells to give a total volume of 200 μL/well. Cells are incubated with the toxins for 6 h, the medium is replaced with fresh culture medium without toxin, and incubation is continued for 48 h. (*see* **Note 12**).

3. Add 50 μL/well of 2.5 mg/mL MTT to the cells, incubate with the cells for 45–120 min at 37°C.

4. Remove the medium. The dark blue oxidized MTT pigment produced by viable cells is dissolved in 100 μL/well of the solvent [0.5% (w/v) SDS, 25 mM HCl, in 90% (v/v) isopropanol] by vortexing the plates, and the oxidized MTT, which is proportional to cell viability, is measured as A_{570} using a microplate reader (*see* **Note 13**).

5. The results of an illustrative experiment are shown in **Fig. 2**. PrAg-U2 efficiently killed the uPA-expressing HeLa cells in a dose-dependent manner, and this cytotoxicity was uPA-dependent because it was blocked by the addition of PAI-1 (*see* **Note 14**).

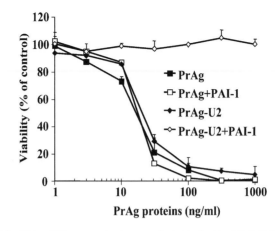

Fig. 2. The cytotoxicity of PrAg-U2 to uPAR-expressing tumor cells is blocked by PAI-1. HeLa cells were cultured to 50% confluence, preincubated with serum-free DMEM containing 100 ng/mL of pro-uPA and 1 µg/mL of Glu-plasminogen with or without 2 µg/mL of PAI-1 for 30 min. Then PrAg and PrAg-U2 combined with FP59 (50 ng/mL) were added to the cells and incubated for 6 h. The toxins were removed and replaced with fresh serum-containing DMEM. MTT was added to determine cell viability at 48 h.

3.4. Activation of PrAg-U2 In Vivo Is Dependent on the Presence of uPA, uPAR, and Plasminogen

The roles of uPAR, plasminogen, and PAI-1 in activation of pro-uPA can be genetically analyzed in vivo by measuring the sensitivity to PrAg-U2/FP59 of mice deficient in these plasminogen activation system components.

1. 6–8-week-old C57BL/6 mice deficient in uPA, uPAR, and plasminogen, and control wild-type mice are injected intraperitoneally with 200µg PrAg-U2 and 10µg FP59 in 500µL PBS (*see* **Note 15**).

2. 6–8-week-old mice deficient in PAI-1 are challenged with various doses of PrAg-U2 (6, 10, 15, and 30 µg) in the presence of 10µg FP59 in 500µL PBS.

3. The mice are monitored closely (checking twice a day) for signs of toxicity for a period of 14 days after injection, by assessing weight loss, inactivity, loss of appetite, inability to groom, ruffling of fur, and shortness of breath. The mice are euthanized by CO_2 inhalation at the onset of obvious malaise.

4. Histological analysis: Mice injected with 200µg PrAg-U2 and 10µg FP59 in PBS are euthanized by CO_2 inhalation at the onset of malaise. The control mice injected with PBS alone are euthanized after 24–36 h by CO_2 inhalation. The mice then are perfused intracardially with cold PBS, followed by 4% paraformaldehyde. The organs are postfixed for 24 h in 4% paraformaldehyde, embedded in paraffin, sectioned, stained with hematoxylin/eosin, and subjected to microscopic analysis by a pathologist unaware of treatment or animal genotype (2–8 mice per treatment group and genotype).

5. Immunostaining of spleen and lymph nodes is performed with a Vectastain ABC peroxidase kit (Vector Laboratories, Inc., Burlingame, CA) with diaminobenzidine as chromogenic

substrate, using rat anti-mouse CD45R/B220 antibodies (Pharmingen, San Diego, CA) to detect B lymphocytes and rabbit anti-human T-cell antibodies (DAKO, Carpinteria, CA) to detect T cells. Apoptotic cells are visualized by terminal deoxynucleotidyltransferase-mediated dUTP nick-end labeling (TUNEL) using an Apotag kit (Intergen, Gaithersburg, MD).

6. The results of an illustrative experiment are shown in **Fig. 3**. All wild-type mice became terminally ill when challenged with

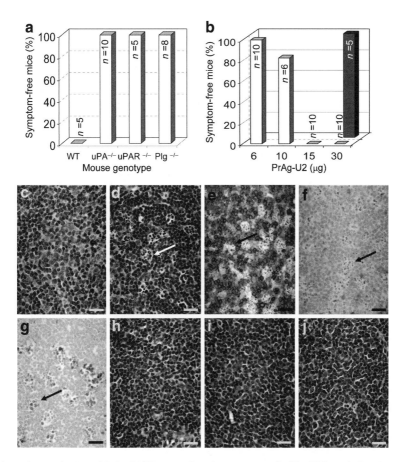

Fig. 3. uPA-dependent activation of PrAg-U2/FP59 requires the presence of uPA, uPAR, and plasminogen in vivo. (**a**) Plg, uPA, and uPAR-deficient mice are hyperresistant to uPA-activated anthrax toxin. Wild-type mice and mice deficient in uPA, uPAR, and Plg were challenged with 200μg PrAg-U2 with 10μg FP59 intraperitoneally and were monitored for disease. All wild-type mice became terminally ill within 24 h of toxin administration, whereas no outward or histological signs of toxicity were detected in uPA, uPAR, and Plg-deficient mice ($P < 0.01$). (**b**) PAI-1-deficient mice are hypersensitive to PrAg-U2. PAI-1$^{-/-}$ (*open bars*) or wild-type control (*solid bars*) mice were challenged with varying concentrations of PrAg-U2 with 10μg FP59 and monitored for disease. All PAI-1$^{-/-}$ mice treated with 15–30μg PrAg-U2 became terminally ill within 24 h of toxin administration, whereas no outward or histological signs of toxicity were detected in wild-type mice challenged with 30μg PrAg-U2 ($P < 0.001$). (**c–j**) Cell-surface uPA-dependent T-cell toxicity of PrAg-U2. Histological appearance of T-cell regions of the spleen of wild type (**c–g**), uPA$^{-/-}$ (**h**), uPAR$^{-/-}$ (**i**), and Plg$^{-/-}$ (**j**) mice 24 h after intraperitoneal injection of PBS (**c**) or 200μg PrAg-U2 with 10μg FP59 (**d–j**). Scattered clusters (examples indicated with arrows) of degenerating lymphocytes in wild-type mice (**d**), absent in PBS-treated wild-type mice (**c**), are identified as subpopulations of T cells, by immunostaining with T-cell (**e**) and B-cell (**f**) antibodies, undergoing apoptosis as visualized by TUNEL-staining (**g**). (**h–j**) shows the absence of T-cell pathology in the spleens of uPA$^{-/-}$ (**h**), uPAR$^{-/-}$ (**i**) and Plg$^{-/-}$ (**j**) mice. (**c**, **d**, and **h–j**) Hematoxylin/eosin staining (bars = 10 μm).

200µg PrAg-U2 with FP59, with cytotoxicity observed in bone marrow, adrenal cortex, osteogenic tissues, T-cell areas of the spleen, and lymph nodes (**Fig. 3c–g**, and data not shown). In contrast, uPA$^{-/-}$, uPAR$^{-/-}$, and Plg$^{-/-}$ mice remained completely healthy, demonstrating that both uPAR and Plg are essential cofactors in the generation of uPA activity in vivo. Microscopic examination of tissues from uPA$^{-/-}$, uPAR$^{-/-}$, and Plg$^{-/-}$ mice challenged with 200µg PrAg-U2 with FP59 failed to demonstrate any signs of cytotoxicity to T-cell areas of the spleen and lymph nodes, bone marrow, adrenal cortex, and osteogenic tissues (**Fig. 3h–j**, and data not shown), providing further evidence that PrAg-U2 is activated by cell surface uPA, and demonstrating that these anatomical locations are principal sites of cell surface uPA activity in vivo. Conversely, PAI-1$^{-/-}$ mice were hypersensitive to PrAg-U2 combined with FP59, with a maximum tolerated dose of about 6 µg (**Fig. 3b**). Microscopic analysis of tissues from PAI-1$^{-/-}$ mice treated with just 20µg PrAg-U2 with FP59 demonstrated bone marrow, T-cell, osteoblast, and adrenal cytotoxicity, similar to wild-type mice treated with much higher concentrations of the engineered toxin (data not shown). All PrAg-U2-treated PAI-1$^{-/-}$ mice also presented profound edema of the small intestine frequently associated with hemorrhaging into the intestinal lumen (data not shown). This condition was never observed in wild-type mice, even when treated with a tenfold higher concentration of the engineered toxin. Taken together, these experiments unequivocally demonstrate that uPA, the binding of uPA to uPAR, and the activation of pro-uPA by plasmin are critical events in the activation of PrAg-U2 in vivo.

4. Notes

1. It is convenient to handle six 3-L flasks of culture each with 500mL enriched FA medium at one time. PrAg and FP59 proteins are secreted into culture medium as major secreted proteins and can usually reach expression levels of 30–50 mg/L.

2. PrAg proteins usually elute at 28% buffer B. Note that culture supernates contain the surface array proteins EA1 and SAP, which have masses like that of PrAg, which sometimes leads to confusion when only SDS gels are used for analysis.

3. At this point, the purity of PrAg proteins usually approaches 95%, with one prominent band evident on gels at the expected molecular mass of 83 kDa.

4. The cell density does not significantly affect the levels of anthrax receptors. Thus, cells near to 100% confluence are usually used for PrAg protein binding and processing analyses.

5. PAI-1 is a major physiological inhibitor of plasminogen activators. Tranexamic acid can compete with plasminogen for cell surface binding sites, and it thus blocks binding of plasminogen to cells. Other inhibitors, such as uPAR blocking antibodies *(21)*, can also be used by preincubation with cells for 30 min before the addition of pro-uPA and toxin.

6. The complete protease inhibitor cocktail tablets are expensive and thus a portion of a tablet can be used as follows: cut one quarter of a tablet using a clean blade and dissolve it in 5–10 mL of RIPA buffer. Precool RIPA lysis buffer on ice before use.

7. After heating, the samples are usually very sticky, and vigorously vortexing to shear the cellular DNA is crucial for successful sample loading on PAGE gel.

8. Thoroughly washing each well of gels using distilled water is crucial to get sharp protein bands.

9. 5% (w/v) Milk (dry milk from Biorad) in TPBS (PBS containing 0.05% Tween 20) is an excellent blocking solution for these antibodies.

10. In the 48-h cytotoxicity assay, cells with initial 30–40% confluency are used to avoid the control untreated wells reaching confluency at 48 h when the data are collected.

11. Other inhibitors, such as aprotinin, α2-antiplasmin, amino-terminal fragment of uPA, or the uPAR blocking antibodies, can also be tested.

12. This cytotoxicity assay can also be performed in regular serum containing DMEM without addition of Glu-plasminogen. In this case, it is not necessary to replace toxin-containing medium with fresh routine culture medium. Similar results can be obtained using either serum-free or serum-containing medium. Fetal bovine serum is a good source for plasminogen.

13. MTT is dissolved in routine cell culture medium.

14. Evidence that the cytotoxicity of PrAg-U2/FP59 to cells is also dependent on cell-surface bound plasminogen and functional uPAR can be found in *(21)*.

15. The maximum tolerated dose of wild-type C57BL/6 mice to PA-U2 is 30 μg in the presence of 10 μg of FP59. The maximum tolerated dose is determined as the highest dose in which outward disease or histological tissue damage is not observed in any mice within a 14-day period of observation.

Acknowledgements

This work was supported by the intramural research program of the National Institute of Allergy and Infectious Diseases and the National Institute of Dental and Craniofacial Research, National Institutes of Health. We thank Andrei Pomerantsev for providing strain BH450, and Rasem Fattah and Dana Hsu for assistance in protein purification.

References

1. Leppla, S.H. (2006) *Bacillus anthracis* toxins. In: J. E. Alouf and M. R. Popoff (eds.), The Comprehensive Sourcebook of Bacterial Protein Toxins. Academic: Burlington, MA, pp. 323–347.

2. Bradley, K.A., Mogridge, J., Mourez, M., Collier, R.J. and Young, J.A. (2001) Identification of the cellular receptor for anthrax toxin. *Nature* 414, 225–229.

3. Scobie, H.M., Rainey, G.J., Bradley, K.A. and Young, J.A. (2003) Human capillary morphogenesis protein 2 functions as an anthrax toxin receptor. *Proc. Natl. Acad. Sci. USA* 100, 5170–5174.

4. Cunningham, K., Lacy, D.B., Mogridge, J. and Collier, R.J. (2002) Mapping the lethal factor and edema factor binding sites on oligomeric anthrax protective antigen. *Proc. Natl. Acad. Sci. USA* 99, 7049–7053.

5. Mogridge, J., Cunningham, K., Lacy, D.B., Mourez, M. and Collier, R.J. (2002) The lethal and edema factors of anthrax toxin bind only to oligomeric forms of the protective antigen. *Proc. Natl. Acad. Sci. USA* 99, 7045–7048.

6. Abrami, L., Liu, S., Cosson, P., Leppla, S.H. and van der Goot, F.G. (2003) Anthrax toxin triggers endocytosis of its receptor via a lipid raft-mediated clathrin-dependent process. *J. Cell Biol.* 160, 321–328.

7. Liu, S. and Leppla, S.H. (2003) Cell surface tumor endothelium marker 8 cytoplasmic tail-independent anthrax toxin binding, proteolytic processing, oligomer formation, and internalization. *J. Biol. Chem.* 278, 5227–5234.

8. Ezzell, J.W., Ivins, B.E. and Leppla, S.H. (1984) Immunoelectrophoretic analysis, toxicity, and kinetics of in vitro production of the protective antigen and lethal factor components of *Bacillus anthracis* toxin. *Infect. Immun.* 45, 761–767.

9. Beall, F.A., Taylor, M.J. and Thorne, C.B. (1962) Rapid lethal effect in rats of a third component found upon fractionating the toxin of *Bacillus anthracis. J. Bacteriol.* 83, 1274–1280.

10. Friedlander, A.M. (1986) Macrophages are sensitive to anthrax lethal toxin through an acid-dependent process. *J. Biol. Chem.* 261, 7123–7126.

11. Duesbery, N.S., Webb, C.P., Leppla, S.H., Gordon, V.M., Klimpel, K.R., Copeland, T.D., Ahn, N.G., Oskarsson, M.K., Fukasawa, K., Paull, K.D. and Vande Woude, G.F. (1998) Proteolytic inactivation of MAP-kinase-kinase by anthrax lethal factor. *Science* 280, 734–737.

12. Vitale, G., Pellizzari, R., Recchi, C., Napolitani, G., Mock, M. and Montecucco, C. (1998) Anthrax lethal factor cleaves the N-terminus of MAPKKs and induces tyrosine/threonine phosphorylation of MAPKs in cultured macrophages. *Biochem. Biophys. Res. Commun.* 248, 706–711.

13. Firoved, A.M., Miller, G.F., Moayeri, M., Kakkar, R., Shen, Y., Wiggins, J.F., McNally, E.M., Tang, W.J. and Leppla, S.H. (2005) *Bacillus anthracis* edema toxin causes extensive tissue lesions and rapid lethality in mice. *Am. J. Pathol.* 167, 1309–1320.

14. Leppla, S.H. (1982) Anthrax toxin edema factor: a bacterial adenylate cyclase that increases cyclic AMP concentrations of eukaryotic cells. *Proc. Natl. Acad. Sci. USA* 79, 3162–3166.

15. O'Brien, J., Friedlander, A., Dreier, T., Ezzell, J. and Leppla, S. (1985) Effects of anthrax toxin components on human neutrophils. *Infect. Immun.* 47, 306–310.

16. Arora, N., Klimpel, K.R., Singh, Y. and Leppla, S.H. (1992) Fusions of anthrax toxin lethal factor to the ADP-ribosylation domain of *Pseudomonas* exotoxin A are potent cytotoxins which are translocated to the cytosol of mammalian cells. *J. Biol. Chem.* 267, 15542–15548.

17. Arora, N. and Leppla, S.H. (1993) Residues 1–254 of anthrax toxin lethal factor are sufficient to cause cellular uptake of fused polypeptides. *J. Biol. Chem.* 268, 3334–3341.

18. Milne, J.C., Blanke, S.R., Hanna, P.C. and Collier, R.J. (1995) Protective antigen-binding domain of anthrax lethal factor mediates

translocation of a heterologous protein fused to its amino- or carboxy- terminus. *Mol. Microbiol.* 15, 661–666.

19. Hobson, J.P., Liu, S., Rono, B., Leppla, S.H. and Bugge, T.H. (2006) Imaging specific cell-surface proteolytic activity in single living cells. *Nat. Methods* 3, 259–261.

20. Liu, S., Netzel-Arnett, S., Birkedal-Hansen, H. and Leppla, S.H. (2000) Tumor cell-selective cytotoxicity of matrix metalloproteinase-activated anthrax toxin. *Cancer Res.* 60, 6061–6067.

21. Liu, S., Bugge, T.H. and Leppla, S.H. (2001) Targeting of tumor cells by cell surface urokinase plasminogen activator-dependent anthrax toxin, *J. Biol. Chem.* 276, 17976–17984.

22. Liu, S., Aaronson, H., Mitola, D.J., Leppla, S.H. and Bugge, T.H. (2003) Potent anti-tumor activity of a urokinase-activated engineered anthrax toxin. *Proc. Natl. Acad. Sci. USA* 100, 657–662.

23. Ke, S.H., Coombs, G.S., Tachias, K., Navre, M., Corey, D.R. and Madison, E.L. (1997) Distinguishing the specificities of closely related proteases. Role of P3 in substrate and inhibitor discrimination between tissue-type plasminogen activator and urokinase. *J. Biol. Chem.* 272, 16603–16609.

24. Ke, S.H., Coombs, G.S., Tachias, K., Corey, D.R. and Madison, E.L. (1997) Optimal subsite occupancy and design of a selective inhibitor of urokinase. *J. Biol. Chem.* 272, 20456–20462.

25. Liu, S., Schubert, R.L., Bugge, T.H. and Leppla, S.H. (2003) Anthrax toxin: structures, functions and tumour targeting. *Expert Opin. Biol. Ther.* 3, 843–853.

26. Singh, Y., Chaudhary, V.K. and Leppla, S.H. (1989) A deleted variant of *Bacillus anthracis* protective antigen is non-toxic and blocks anthrax toxin action *in vivo. J. Biol. Chem.* 264, 19103–19107.

27. Liu, S., Leung, H.J. and Leppla, S.H. (2007) Characterization of the interaction between anthrax toxin and its cellular receptors. *Cell. Microbiol.* 9, 977–987.

28. Pomerantsev, A.P., Sitaraman, R., Galloway, C.R., Kivovich, V. and Leppla, S.H. (2006) Genome engineering in *Bacillus anthracis* using Cre recombinase. *Infect. Immun.* 74, 682–693.

29. Carmeliet, P., Schoonjans, L., Kieckens, L., Ream, B., Degen, J., Bronson, R., De, V.R., van den Oord, J.J., Collen, D. and Mulligan, R.C. (1994) Physiological consequences of loss of plasminogen activator gene function in mice. *Nature* 368, 419–424.

30. Bugge, T.H., Suh, T.T., Flick, M.J., Daugherty, C.C., Romer, J., Solberg, H., Ellis, V.,

Dano, K. and Degen, J.L. (1995) The receptor for urokinase-type plasminogen activator is not essential for mouse development or fertility. *J. Biol. Chem.* 270, 16886–16894.

31. Bugge, T.H., Flick, M.J., Daugherty, C.C. and Degen, J.L. (1995) Plasminogen deficiency causes severe thrombosis but is compatible with development and reproduction. *Genes Dev.* 9, 794–807.

32. Carmeliet, P., Kieckens, L., Schoonjans, L., Ream, B., van Nuffelen, A., Prendergast, G., Cole, M., Bronson, R., Collen, D. and Mulligan, R.C. (1993) Plasminogen activator inhibitor-1 gene-deficient mice. I. Generation by homologous recombination and characterization. *J. Clin. Invest.* 92, 2746–2755.

33. Andreasen, P.A., Kjoller, L., Christensen, L. and Duffy, M.J. (1997) The urokinase-type plasminogen activator system in cancer metastasis: a review. *Int. J. Cancer* 72, 1–22.

34. Dano, K., Romer, J., Nielsen, B.S., Bjorn, S., Pyke, C., Rygaard, J. and Lund, L.R. (1999) Cancer invasion and tissue remodeling--cooperation of protease systems and cell types. *APMIS* 107, 120–127.

35. Romer, J., Bugge, T.H., Pyke, C., Lund, L.R., Flick, M.J., Degen, J.L. and Dano, K. (1996) Impaired wound healing in mice with a disrupted plasminogen gene. *Nat. Med.* 2, 287–292.

36. Lund, L.R., Bjorn, S.F., Sternlicht, M.D., Nielsen, B.S., Solberg, H., Usher, P.A., Osterby, R., Christensen, I.J., Stephens, R.W., Bugge, T.H., Dano, K. and Werb, Z. (2000) Lactational competence and involution of the mouse mammary gland require plasminogen. *Development* 127, 4481–4492.

37. Heymans, S., Luttun, A., Nuyens, D., Theilmeier, G., Creemers, E., Moons, L., Dyspersin, G.D., Cleutjens, J.P., Shipley, M., Angellilo, A., Levi, M., Nube, O., Baker, A., Keshet, E., Lupu, F., Herbert, J.M., Smits, J.F., Shapiro, S.D., Baes, M., Borgers, M., Collen, D., Daemen, M.J. and Carmeliet, P. (1999) Inhibition of plasminogen activators or matrix metalloproteinases prevents cardiac rupture but impairs therapeutic angiogenesis and causes cardiac failure. *Nat. Med.* 5, 1135–1142.

38. Nielsen, L.S., Hansen, J.G., Skriver, L., Wilson, E.L., Kaltoft, K., Zeuthen, J. and Dano, K. (1982) Purification of zymogen to plasminogen activator from human glioblastoma cells by affinity chromatography with monoclonal antibody. *Biochemistry* 21, 6410–6415.

39. Bugge, T.H., Flick, M.J., Danton, M.J., Daugherty, C.C., Romer, J., Dano, K., Carmeliet, P., Collen, D. and Degen, J.L.

(1996) Urokinase-type plasminogen activator is effective in fibrin clearance in the absence of its receptor or tissue-type plasminogen activator. *Proc. Natl. Acad. Sci. USA* 93, 5899–5904.

40. Carmeliet, P., Moons, L., Dewerchin, M., Rosenberg, S., Herbert, J.M., Lupu, F. and Collen, D. (1998) Receptor-independent role of urokinase-type plasminogen activator in pericellular plasmin and matrix metalloproteinase proteolysis during vascular wound healing in mice. *J. Cell Biol.* 140, 233–245.

Chapter 11

On-Demand Cleavable Linkers for Radioimmunotherapy

Pappanaicken R. Kumaresan, Juntao Luo, and Kit S. Lam

Summary

Radioimmunotherapy (RIT) using radiolabeled antibodies or its fragments holds great promise for cancer therapy. However, its clinical potential is often limited by the undesirable radiation exposure to normal organs such as liver, kidney, and bone marrow. It is important to develop new strategies in RIT that enable protection of vital organs from radiation exposure while maintaining therapeutic radiation dose to the cancer. One way to achieve this is to clear radiometal rapidly from the circulation after accumulation of radioimmunoconjugates (RIC) in the tumor. Our strategy is to place a highly efficient and specific cleavable linker between radiometal chelate and the tumor targeting agent. Such linker must be resistant to cleavage by enzymes present in the plasma and tumor. After radiotargeting agents have accumulated in the tumor, a cleaving agent (protease) can be administered to the patient "on demand" to cleave the specific linker, resulting in the release of radiometal from the circulating RIC, in a form that can be cleared rapidly by the kidneys. TNKase®, a serine protease tissue plasminogen activator and thrombolytic agent, which has been approved for clinical use in patient with acute myocardial infarction, was selected as an on-demand cleaving agent in our model. TNKase® specific on-demand cleavable (ODC) linkers were identified through screening random internally quenched fluorescent resonance energy transfer (FRET) "one-bead-one-compound" (OBOC) combinatorial peptide libraries. FRET-OBOC peptide libraries containing L-amino acid(s) in the center of the random linear peptide and D-amino acids flanking both sides of the L-amino acid(s) were used for screening. Peptide beads susceptible to TNKase® but resistant to plasma and tumor-associated protease cleavage were isolated for sequence analysis. The focus of this chapter is on the methods that have been used to identify and characterize ODC linkers and protease-specific substrates.

Key words: Protease substrates, On-demand cleavable linkers, Radioimmunotherapy, FRET-OBOC libraries, Combinatorial chemistry.

1. Introduction

Radioimmunotherapy (RIT), while promising, is limited by the undesirable radiation effects to normal organs *(1–3)*. One of the strategies is minimizing the radiation effects to normal organs

Toni M. Antalis and Thomas H. Bugge (eds.), *Methods in Molecular Biology, Proteases and Cancer, Vol. 539*
© Humana Press, a part of Springer Science+Business Media, LLC 2009
DOI: 10.1007/978-1-60327-003-8_11

by placing highly efficient and specific cleavable linkers between radiometal chelates and the tumor targeting agents *(4–6)*. Such linkers must be resistant to cleavage by enzymes present in the plasma and tumor. After radiotargeting agents have accumulated in the tumor, a cleaving agent (protease) can be administered "on demand" to cleave a specific linker, resulting in the release of radiometal from the circulating radioimmunoconjugate (RIC) in a form that will have rapid renal clearance through urine (**Fig. 1**). Such an on-demand cleavable (ODC) linker can be discovered through screening one-bead-one-compound (OBOC) combinatorial peptide libraries. Here, we describe the FRET-OBOC combinatorial library synthesis, screening methods, and applications. For other applications about OBOC combinatorial library methods, please refer to our other publications *(7–10)*.

1.1. Fmoc-Protected Synthetic Chemistry for Solid-Phase Peptide Synthesis

Compared with solution-phase peptide synthesis, solid-phase peptide synthesis has obvious advantages: easy separation of the peptide intermediate from the soluble coupling reagents and the unreacted reagents by simple filtration and washing of the solid supports, and complete coupling on supports driven by excess of reagents. In addition, most of these operations are amenable to automation. During the peptide synthesis, the N terminus and the reactive side chain groups of amino acids, such as OH, SH, NH_2, COOH, etc., need to be protected using orthogonal cleavable groups. 9-Fluorenylmethyloxy carbonyl (Fmoc) is a basic cleavable protecting group for amine and is stable in acidic and peptide coupling conditions. Fmoc-based solid-phase peptide synthesis is widely applied *(11, 12)* and illustrated in **Fig. 2**. Resins with an acidic labile linker, such as Wang resin and Rink amide resin, and Fmoc-protected amino acids with acid-labile side chain protecting group are used in peptide synthesis. There are three steps in solid peptide synthesis: (a) coupling, (b) deprotection, and (c) cleavage. In between these steps, sufficient wash is required to remove the reagents and impurities. In the coupling step, peptide

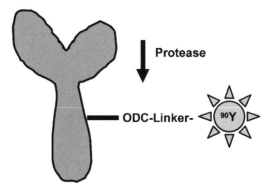

Fig. 1. Schematic of on-demand cleavable linkers (ODC).

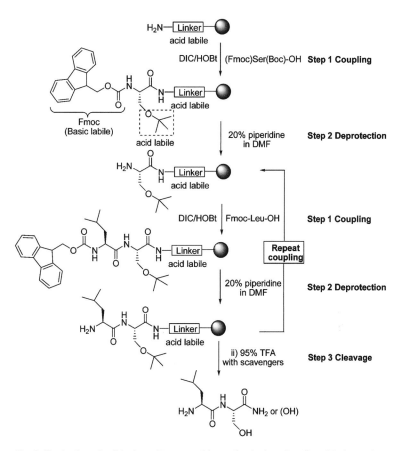

Fig. 2. Illustration of solid-phase Fmoc peptide synthesis: Leu-Ser dipeptide is used as an example.

bond is formed between carboxyl group of N-terminal (Fmoc)-protected amino acid in solution phase and the free amino group of the amino acids attached onto the resin by liberating a water molecule. This process is driven to complete by adding excess of protected amino acids in presence of coupling reagents, such as HOBt (1-hydroxybenzotriazole) and DIC (diisopropylcarbodiimide). It is important to note that in chemical peptide synthesis, peptide bond formation starts from carboxyl terminal that means that the last amino acid in the sequence needs to couple first to the resin and the first amino acid in the sequence comes for the last coupling. After last amino acid coupling, the peptide is released from peptide by the treatment of peptide resin with trifluoroacetic acid (TFA), and at the same time, the side chain protecting groups are removed simultaneously. The crude peptide is precipitated in cold ether after most of the TFA is removed (with a gentle stream of nitrogen), and the precipitate is washed three times with cold ether, followed by the purification on reverse phase HPLC.

1.2. Fluorescence Resonance Energy Transfer (FRET)

FRET (Fluorescence Resonance Energy Transfer) is a technique widely used in proteomics for measuring interactions between two biomolecules (13–15). In regular (non-FRET) fluorescence, the fluorescent molecule absorbs electromagnetic energy at a particular wavelength (the excitation frequency) and emits that energy at a different wavelength (the emission frequency). In other words, each fluorophore has a two-peaked spectrum – the first peak is the excitation peak and the second is the emission peak. For FRET effect, the emission peak of the donor must overlap with the excitation peak of the acceptor. In FRET, the added light energy to the donor at its excitation frequency is transferred to the acceptor and the acceptor reemits the light energy at its own emission wavelength. The net result is that the donor emits less energy than it normally emits without the acceptor. Some of the FRET donor and acceptor pairs have been shown in **Table 1**. FRET occurs only when the donor and the acceptor are situated within 20–100Å of each other for better energy transfer.

1.3. FRET-OBOC Combinatorial Libraries

Fluorescent resonance energy transfer (FRET)-quenched random OBOC combinatorial peptide libraries can be synthesized by incorporating fluorophor (donor) at the carboxyl terminal and quencher (acceptor) at the amino terminal of a peptide using Fmoc chemistry and split-mix synthesis method (16, 17). In this method, the combinatorial chemistry libraries are synthesized on 100–300-μm size bead resins. Because of its higher porosity due to high swelling properties, polyethylene glycol-*poly*-(*N*,*N*-dimethylacrylamide) copolymer (PEGA) bead resin is more suitable for protease substrate screening than less porous TentaGel resin. The porosity of the PEGA beads allows the protease to penetrate deeper into the interior of the beads, which produces a favorable fluorescent signal-to-noise ratio.

In this library, the first amino acid proximal to the solid support has a free amino group on the side chain (e.g., lysine) to which a fluorescent (donor) molecule is coupled. The last amino acid (distal to the solid support or N-terminus) in this library is

Table 1
Commonly used fluorescence and quencher pair

Fluorophor (donor)	Quencher (acceptor)
Ortho-aminobenzoic acid	Phe(4-NO$_2$), nitrobenzamide, 2,4-dinitrophenyl (DNP), Tyr(3-NO$_2$), 4-nitroaniline
Aminocoumarin, aminoquinolinone	Nitro-aromatic derivatives
5-((2-Aminoethyl)-amino)napthelensulfonic acid (EDANS)	4-((4′-(Dimethylamino)phenyl)azo)-benzoic acid (DABCYL)
Tryptophan	Dansyl (Dns)

a quencher (acceptor). We have selected 2-amino benzoic acid (fluorescent label) and 3-nitro tyrosine (quencher) as fluorescent/quencher pair to synthesize the FRET-OBOC library on PEGA resin using a standard "split-mix" approach. If a significant number of false-positive fluorescent beads remain after the library synthesis, the library may be capped with a different quencher (**Table** 1) *(10, 18)*. This capping step should be performed prior to side chain deprotection. After fluorophor coupling in the first position, beads are split into number of small columns as determined by the number of amino acids used in the second position. The beads in each column are coupled to one type of amino acid (e.g., first column is for "alanine," second one is for "methionine," and so on). After coupling, beads from all the columns are mixed and split into number of small columns as determined by the number of amino acids used in the third position. Examples of FRET-OBOC libraries are Y(NO$_2$)-xxxXxxx-K(Abz), Y(NO$_2$)-xxxXXxx-K(Abz), and Y(NO$_2$)-xxXXXxx-K(Abz), wherein Y(NO$_2$) represents 3-nitrotyrosine, K(Abz) represents lysine bearing a fluorescent 2-aminobenzoyl group on its ε-amino group, X represents one of 19 natural L-amino acids excluding cysteine, and x represents one of the D-isomers of 18 natural L-amino acids plus glycine. Theoretically, there are $19^7 = 8.9 \times 10^8$ possible permutations of peptide sequences in these libraries (**Fig. 3**). The use of D-amino acids as flanking residues in the library is critically important for substrate specificities because peptide bonds between these D-amino acids are totally resistant to proteolysis, and therefore will not be cleaved by the numerous proteases present in the blood. However, these D-amino acids will interact with the enzyme active site and will contribute to substrate specificity. The peptide bonds that are susceptible to proteolytic cleavage are only those with an L-amino acid at the P1 position.

1.4. On-Demand Cleaving Agents and Cleavable Linkers

We chose TNKase®, a tissue plasminogen activator, as the cleavage agent because it has already been in clinical use for the treatment of myocardial infarction, and is therefore readily available *(19–21)*. One of the first TNKase® specific substrates, rqYKYkf, identified through library screening Y(NO$_2$)-xxXXXxx-K(Abz), was used as an ODC linker to conjugate DOTA (a metal chelate) to ChL6 (a monoclonal antibody known to target breast cancer) *(4)*. This antibody conjugate was stable in plasma for 7 days while preserving the immunoreactivity to intact tumor cells. The addition of TNKase® at clinically achievable levels (10 μg/mL) resulted in the release of 28% of the radiometal from the radioimmunoconjugate within 72 h. More recently, we have developed additional ODC linkers that are equally specific but much more efficient, with over 80% of the radiometal released from the immunoconjugate 72 h after treatment with TNKase (10 μg/mL). Work is currently underway in our laboratory to develop ODC linkers that are even more efficient.

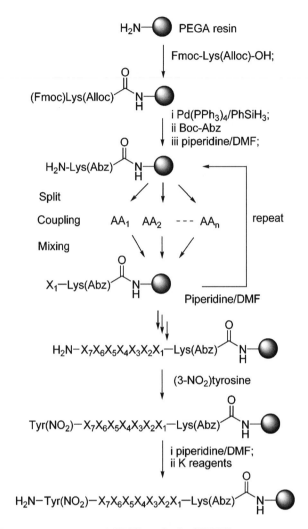

Fig. 3. One-bead-one-compound (OBOC) synthesis of FRET library.

In this chapter, we describe the experimental methods to synthesize and screen the FRET-OBOC libraries, the synthesis of FRET-peptides for enzyme kinetic studies and for antibody conjugation, and finally the in vitro evaluation of the radioimmunoconjugate containing an ODC linker.

2. Materials

2.1. Preparation of "Fluorescent-Quenched" OBOC Combinatorial Peptide Library

1. PL-PEGA resin (0.4 mmol/g) 150–300 μm (cat# 1432-4879) was purchased from Polymer laboratories, USA.

2. Boc-2-Abz-OH and Boc-Tyr(3-NO$_2$)-OH was purchased from ChemImpex (Wood Dale, IL).

3. HOBt (1-hydroxybenzotriazole) was purchased from GL Biochem (Shanghai, China).

4. DIC (diisopropylcarbodiimide) was purchased from Advanced ChemTech (Louisville, KY). DMF (N,N-dimethylformamide) was purchased from VWR (Brisbane, CA).

5. DCM (dichloromethane), MeOH (methanol), diethyl ether, and acetonitrile were purchased from Fisher (Houston, TX). All solvents were used directly for library synthesis without any purification unless otherwise noted.

6. All other chemical reagents were purchased from Aldrich (Milwaukee, WI).

2.2. FRET Combinatorial Library Synthesis

1. Prepare all the α-Fmoc-protected amino acids (N-fluorenyl-methoxycarbonyl α-amino acids) dissolved in DMF (N,N-dimethylformamide) containing equimolar amount of HOBt (1-hydroxybenzotriazole) just before use. The amino acids can be stored at 4°C for a week.

2. Add threefold excess of N^α-Fmoc-AAs, HOBt, and DIC for each coupling step to ensure completion of the coupling reaction.

3. Use the following equation to calculate required amount of HOBt (1-hydroxybenzotriazole mono hydrate). The amount of DMF mentioned here is adjusted to 100 mg of resin per coupling.

$$W_{\mathrm{HOB}} = (153.2)(3S)(g)(N),$$

where W_{HOBt} is the required weight of HOBt in mg, 153.2 is the molecular weight of HOBt, S is the loading substitution of the resin in mmol/g, g is the mass of the resin in gram, and N is the number of couplings involved in the synthesis.

4. Dissolve HOBt with DMF (use 1 mL of DMF per coupling for 100-mg PEGA resin).

5. Use the following equation to calculate required amount of amino acid per coupling.

$$W_{\mathrm{aa}} = (\mathrm{MW}_{\mathrm{aa}})(3S)(g),$$

where W_{aa} is the mass (in mg) of each protected amino acid required for the library synthesis, $\mathrm{MW}_{\mathrm{aa}}$ is the molecular weight of the protected amino acid, g is the mass of the resin in gram, and S is the loading substitution of the resin in mmol/g. Dissolve amino acid weighed for one coupling in 1 mL of HOBt dissolved DMF.

6. Use the following equation to calculate required amount of DIC per coupling.

$$V_{\mathrm{DIC}} = (126.2)(3S)(g)/(0.806),$$

where V_{DIC} is the volume (in µL) of DIC required for each amino acid coupling reaction, 126.2 is the molecular weight of DIC, 0.806 is the density of DIC, S is the loading substitution of the resin in mmol/g, and g is the mass of the resin in gram.

7. Cleavage cocktail with the following composition was freshly made each time: Trifluoro acetic acid (82.5 mL), phenol (7.5 g), thioanisole (5.0 mL), water (2.5 mL), and triisopropylsilane (2.5 mL).

2.3. Kaiser Test (22)

1. Ninhydrin solution: 5% in ethanol; store in brown dropper bottle for longtime usage.

2. Crystalline phenol solution: 80 g in 20 mL ethanol in a brown dropper bottle.

3. Pyridine/KCN solution: 2 mL of 0.001 M aqueous solution of KCN (potassium cyanide) with 98 mL of pyridine.

4. To perform the test, transfer a few resin beads to a 6 mm × 50 mm glass test tube (disposable culture tubes). Wash the beads with ethanol and add a drop of the aforementioned three testing solutions. Heat the tube to 100°C for 5 min in a heating block. Blue (brown for proline) colored beads and solution (positive test) indicate the presence of free amino groups on the resin. Colorless beads (negative test) and yellow solution indicate complete coupling. The quality of pyridine is important; amino impurities in pyridine may result in a false-positive Kaiser test.

2.4. Chloranil Test (23)

1. Reagents: 2% acetaldehyde in DMF (v/v) and 2% *p*-chloranil in DMF (w/v).

2. Transfer a few resin beads to a small glass test tube and wash with DMF.

3. Add one drop of each of the two reagents and keep at room temperature for 5 min.

4. Appearance of blue or dark green color indicates the presence of free amino groups on the resin. Colorless beads indicate complete coupling.

2.5. Preparation of Ketone Linker (24)

1. Add 1.24 mL of levulinic acid (10 mmol) and 1.16-g *N*-hydroxyl succinimide (HOSu) into a single neck flask; and 30 mL of DMF/DCM (1:5 v/v) mixture solution is charged into the flask.

2. Stir the mixture solution with a magnetic stir while cooling the reaction mixture in an ice bath for 10 min.

3. Add dropwise to the reaction mixture a solution of 2.06 g (10 mmol) of dicyclohexylcarbodiimide (DCC) in 10-mL DCM. Then, agitate the reaction at 4°C overnight.

4. Filter the reaction mixture and wash it with water in an extraction funnel, collect the organic layer, and dry it over anhydrous Mg_2SO_4 for 4 h.

5. Remove the drying agents by filtration and remove the organic solvent by evaporation under vacuum.

6. Add 20 mL of cold ether into the residue and collect the precipitated white product by filtration.

3. Methods

3.1. Synthesis of FRET-OBOC Combinatorial Libraries

Here, we describe the synthesis of a heptamer-FRET-OBOC library tagged with 2-aminobenzoic acid (fluorochrome) at C-terminus and 2-nitrotyrosine (quencher) at N-terminus (10).

1. Soak 2 g of PL-PEGA resin (0.4 mmol/g) 150–300 µm (cat# 1432-4879) in a 50-mL polypropylene column (Pierce) overnight with DMF.

2. Drain the DMF and add Fmoc-Lys(Alloc)-OH (1.086 g, 2.4 mmol), HOBt (0.367 g, 2.4 mmol), and DIC (375.8 µL, 2.4 mmol) in DMF (20 mL) to the resin. Remove very small amount of beads after 2–3 h of mixing and perform Kaiser test as mentioned in **Subheading 2.3**. Agitate the resulting mixture until the ninhydrin test becomes negative.

3. Wash the resin 3× with DMF, 3× with MeOH (methanol), and 3× with DCM.

4. To remove Alloc protecting group, add phenylsilane solution ($PhSiH_3$) (1.87 mL, 16.0 mmol) in DCM (10 mL) followed by tetrakis(triphenyl phosphine) palladium(0) solution ($Pd(PPh_3)_4$) (221.7 mg, 0.192 mmol) in DCM (2.2 mL). Shake the mixture in the presence of argon atmosphere for 30 min. Because there is gas generated in this process, release pressure from time to time by opening the cap of the reaction vessel. It is very important to release the gas to avoid gas pressure related explosion. Drain the solution and repeat this process (**step 4**).

5. Remove the supernatant and wash the resin thoroughly with DCM, DMF, MeOH, and DMF (**step 3**).

6. To the resin, add a solution of Boc-2-Abz-OH (379.2 mg, 1.6 mmol), HATU (608 mg, 1.6 mmol), and DIEA (557.8 µL, 3.2 mmol) in DMF (20 mL) and agitate the mixture until the ninhydrin test becomes negative.

7. Drain the solution and add 20% piperidine solution in DMF (20 mL) to the resin. Agitate the mixture for 15 min, and drain the supernatant. Repeat this process again and wash the resin thoroughly 3× with DMF, MeOH, and DMF.

8. Evenly distribute the beads into 20 polypropylene columns (2 mL), each with a tight-sealing cap, a frit, and a stopper, and couple on to the α-amino group of the lysine using standard Fmoc chemistry *(12)*.

9. Add 1.0 mL of prepared D or L-amino acid/HOBt solution and 12.7 µL of DIC to each column. Agitate the reaction mixture for 1 h at room temperature and monitor by performing Kaiser test. For those columns with incomplete coupling reactions, prolong the reaction time or a "repeat coupling" by adding fresh amino acid/HOBt solution and DIC until Kaiser test becomes negative.

10. After first coupling, combine all the resin in a siliconized glass column or polypropylene column with a frit at the bottom. Wash the resin with DMF three (15 mL each wash) times. Remove $N\alpha$-Fmoc protecting group by incubating twice in 10 mL of 20% piperidine/DMF on a rotator for 15 min each.

11. Wash the resin thoroughly 3× with DMF, MeOH, and DMF, respectively.

12. Remove very small amounts of beads for Kaiser test to confirm the Fmoc deprotection (deep blue color).

13. Split the beads 100 mg per column into 20 minicolumns with 2-mL capacity.

14. Couple one amino acid solution per column; the use of either D-amino acid or L-amino acid depends on the designed structure of the amino acid.

15. Repeat the cycle of resin distribution, amino acid coupling, Kaiser test, resin mixing, resin washing, Fmoc deprotection, resin washing, and Kaiser test **(steps 7–12)** until last amino acid assembly is completed.

16. Couple Boc-Tyr(3-NO$_2$)-OH as the last amino acid to quench the fluorescence of the beads.

17. After the final coupling, wash the resin with DMF, MeOH, and DCM, and then dry the resin under vacuum for 2 h.

18. In the last coupling, Fmoc-Tyr(3-NO$_2$)-OH can be used instead of Boc-Tyr(3-NO$_2$)-OH. In that case, the $N\alpha$-Fmoc protecting group on the last amino acid must be removed prior to side-chain deprotection (**step 10**).

19. Wash the resin with DCM (15 mL × 3), and dry under vacuum for 2 h.

20. Remove the side-chain protecting groups by agitating the resin with 30-mL cleavage solution for 4 h at room temperature.

21. Remove the supernatant and wash the resin with DMF (15 mL × 5), methanol (15 mL × 3), DCM (15 mL × 3), DMF

(15 mL × 3), 30% water/DMF (15 mL × 1), 60% water/DMF (15 mL × 1), water (15 mL × 1), and PBS buffer (15 mL × 10).

22. Store the FRET-OBOC peptide library in 0.05% sodium azide/PBS at 4°C.

3.2. Screening FRET-OBOC Combinatorial Libraries (4, 7)

1. Select FRET-OBOC peptide libraries and wash several times with PBS at room temperature to remove all the preservatives. For initial screening, select 1 mL of bead volume (~350,000 beads) in 2-mL polypropylene column.

2. Equilibrate the library beads with appropriate enzyme buffer for 3 h.

3. Remove unquenched beads before adding enzyme. Most of the quenched FRET libraries have unquenched beads (up to 1%). It can be removed either by passage through fluorescent bead sorter (COPAS™ SELECT, Union Biometrica) or manual picking under a fluorescent microscope. Picking false-positive beads manually is a time-consuming process whereas the bead sorter can sort up to 30,000 beads per hour (**Notes 4.1** and **4.2**).

4. Add the appropriate units (determined by purpose of the study) of protease to assay buffer and incubate the beads under assay conditions recommended for that particular protease.

5. Invert or rotate the beads frequently to facilitate contact between the beads and enzyme.

6. After required time of incubation (usually 1 h) drain the buffer and incubate the beads with 8 M guanidine hydrochloride.

7. Wash the beads several times with water and transfer to a Petri dish.

8. Look for the fluorescent beads under excitation/emission range 360/420 nm.

9. Positive beads show fluorescence stronger at the periphery and weaker at the center (donut shape) (**Fig. 4**). It is important to optimize the conditions so that the selected positive beads are not overly digested. Otherwise only partial sequence can be recovered.

10. Pick the positive beads manually or by bead sorter.

3.3. Preparation of the Positive Beads for Microsequencing

1. Incubate the beads with 8 M guanidine hydrochloride for 1 h with gentle agitation in a 6-mm Petri dish or in a 6-well plate.

2. Transfer the beads manually under light microscope to wells containing water.

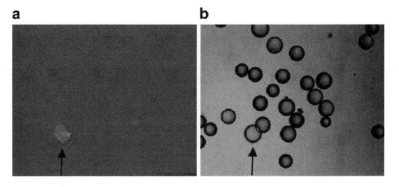

Fig. 4. Screening FRET-OBOC combinatorial peptide library with TNKase at clinically achievable level (10 µg/mL). TNKase-susceptible fluorescent bead is shown by arrow in (**a**) and visible spectrum of the same field is shown in (**b**).

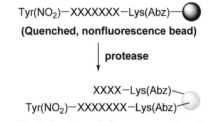

Fig. 5. Schematic of linear FRET-OBOC combinatorial library structure before and after digestion.

3. Wash the beads with water two times to completely remove the guanidine HCl.

4. Carefully transfer a single bead onto a glass filter provided by the manufacturer of the sequencer.

3.4. Sequencing Positive Beads

1. Sequence the beads with a microsequencer (Procise, ABI) to identify substrate sequences.

2. Some of the peptides on the positive beads might be cleaved at the proteolytic site and as a result, two or more PTH-amino acids can be detected in each cycle of Edman degradation using the microsequencer *(25)*.

3. This might complicate the sequence analysis somewhat; at the same time, it also allows us to determine the proteolytic cleavage site of the peptide, in addition to its uncleaved sequence (**Fig. 5**).

3.5. Synthesis of Individual Fluorescent-Quenched ODC Linkers

It is necessary to have the products in solution form without tagging to beads for doing enzyme kinetic studies. After successful identification of protease-specific substrates these can be synthesized by Fmoc chemistry as mentioned in library synthesis with slight modification.

1. Add rink amide MBHA resin (200 mg, 0.09 mmol) to a 5-mL column and soak overnight with DMF.

2. Drain the supernatant and add 20% piperidine solution in DMF (2 mL).

3. Gently agitate the mixture for 15 min and drain the solution, and then repeat this process.

4. Wash the resin thoroughly 3× with DMF, MeOH, and DMF.

5. Add Fmoc-Lys(Alloc)-OH (122.2 mg, 0.27 mmol), HOBt (36.5 mg, 0.27 mmol), and DIC (42.3 μL, 0.27 mmol) in DMF (2 mL) to the resin.

6. Agitate the mixture in a rotator until coupling is complete as confirmed by a negative ninhydrin test (Kaiser test) on a minute bead sample as mentioned in **Subheading 2.3**.

7. Wash the resin 3× with DMF, MeOH, and DCM.

8. Add a solution of $PhSiH_3$ (222 μL, 2.16 mmol) in DCM (1 mL) to the resin, followed by a solution of $Pd(PPh_3)_4$ (25.0 mg, 0.0216 mmol) in DCM (2.2 mL), and agitate mixture in argon atmosphere for 30 min to remove the Nε-Alloc group of lysine (refer **Subheading 3.1, step 4**).

9. Repeat this process.

10. Remove the supernatant and wash the resin thoroughly 3× with DCM, DMF, MeOH, and DMF.

11. Add a solution of Boc-2-Abz-OH (64.1 mg, 0.27 mmol), HOBt (36.5 mg, 0.27 mmol), and DIC (42.3 μL, 0.27 mmol) in DMF (2 mL), and agitate until the ninhydrin test is negative.

12. Remove the supernatant and wash with DMF, and remove α-Fmoc protecting group of the lysine by piperidine treatment as described earlier (**steps 2–5**).

13. Synthesize the peptide by coupling the amino acids starting from carboxyl terminal first on the α-amino group of the lysine using standard Fmoc chemistry (26).

14. Couple Boc-Tyr($3-NO_2$)-OH as the last amino acid.

15. After coupling all the amino acids in the peptide sequence, wash the resin with DMF, MeOH, and DCM, and then dry in vacuo.

16. Add 4 mL of cleavage mixture to the dried resin at ice-bath temperature.

17. Warm the mixture slowly to room temperature and allow it to mix for 6 h in a rotator.

18. Collect the supernatant and wash the resin with TFA (3 × 1 mL). Then, combine both supernatants and concentrate to 1 mL under a stream of nitrogen.

19. Precipitate the peptide by dilution with ice-chilled ethyl ether (10 mL) and centrifuge at $500 \times g$ for 10 min.

20. Wash the peptide pellet with ice-chilled ethyl ether 3×, and dry in vacuo.

21. Analyze and purify the crude peptide by reverse phase HPLC and characterize by MALDI-TOF MS.

3.6. Stability and Specificity of ODC Linkers

It is important to analyze the stability of the linkers in the biological fluids such as plasma and tissue culture supernatants (**Fig. 6**).

1. Incubate 100 nmol/mL of fluorescent-quenched ODC linkers in PBS containing 10% human plasma at 37°C.

2. The stability of the ODC linkers can be monitored over time using a fluorescence plate reader (Tecan®) at 360/410 nm excitation/emission spectrum. Susceptible ODC linkers will show increased fluorescence, while stable linkers will not.

3. Incubate the ODC linkers over a period of 12 h.

4. Select the ODC linkers that are stable in plasma (overnight incubation) for subsequent cancer cell supernatant stability analysis.

5. For this, collect cell culture supernatant from several cancer cell lines (approximately 1–5 million cells/mL) that were passed one day before the collection.

6. Incubate 100 nmol/mL of the ODC linkers with cancer cell culture supernatant and monitor its susceptibility in fluorescence plate reader at 360/410 nm excitation/emission spectrum.

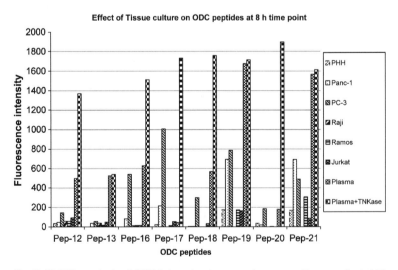

Fig. 6. Stability analysis of ODC linkers in plasma and cancer cell supernatant: 100 nmol/mL of ODC linkers incubated with human plasma (dilute 50% with PBS) and various cancer cell supernatants for 8 h. Susceptibility was shown as measured fluorescent intensity in Y-axis and a list of cancer cells is shown in the right side. Pep-20 showed high stability in various cancer cell culture supernatant and plasma; at the same time, it is highly susceptible to TNKase enzyme.

7. Select the ODC linkers that show stability in both plasma and various cancer cell culture supernatants for enzyme kinetic assays.

3.7. Enzyme Kinetics of ODC Linker Cleavage

1. Prepare ODC substrates at various concentrations ranging from 1 µM to 100 mM in an enzyme assay buffer.

2. Aliquot 100 µL of various concentrations of ODC substrates in a 96-well plate in triplicate.

3. Add 1 unit of enzyme (arbitrary and various to enzyme to enzyme) to all the wells and incubate at 37°C for 1 h with occasional shaking. The amount of enzyme addition and incubation time can be adjusted to the type of proteases used.

4. Measure fluorescence at 360/410 nm excitation/emission spectrum using Tecan® fluorescence plate reader.

5. Determine the K_m value of each substrate by plotting the reaction cleavage rate as observed fluorescence units in the Y-axis and substrate concentrations in the X-axis.

6. Calculate Michaelis constant K_m of the substrate by intersecting substrate concentration from the graph at the half of the maximal velocity (V_{max}) (**Fig. 7**).

7. Alternatively statistical software such as "Graph pad prism4" may be used.

3.8. Synthesis of Biotin- or DOTA-Derivatized ODC Linker with an Aminooxyacetyl (Aoa) Group

Conjugation chemistry such as thiol linkage, keto–oxime bond, and click chemistry can be used to conjugate the ODC linkers to antibody *(27, 28)*. Select a method based on its applications. Here, we describe a conjugation method using keto–oxime bond. The tagged ODC linkers can be synthesized in two forms, biotin-spacer-rqYKYkf-spacer-K(Aoa)-NH$_2$ and DOTA-spacer-rqYKYkf-spacer-K(Aoa)-NH$_2$, using a procedure similar to the synthesis of fluorescence-quenched peptides (**Subheading 3.5, steps 1–10**.

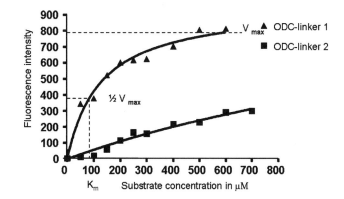

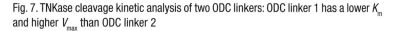

Fig. 7. TNKase cleavage kinetic analysis of two ODC linkers: ODC linker 1 has a lower K_m and higher V_{max} than ODC linker 2

After the Alloc deprotection, instead of Boc-2-Abz-OH, add Boc-Aoa-OH to couple with ε-amino group of the lysine.

1. Remove α-Fmoc protecting group by piperidine treatment as described earlier.

2. Add a linear hydrophilic spacer to couple to the α-amino group of the lysine as a spacer using our previously published procedure (24).

3. Couple ODC peptide as described in **Subheading 3.5**. After the coupling of the last amino acid, remove N-terminus Fmoc protection group by piperidine treatment.

4. Add an additional spacer between the DOTA and the peptide.

5. Remove N-terminus Fmoc protection group by piperidine treatment.

6. Add a solution of DOTA-NHS-ester (112.0 mg, 0.135 mmol) and DIEA (156.8 μL, 0.9 mmol) to the resin and couple until the ninhydrin test becomes negative.

7. Release the peptides from the resin and purify using the procedure described earlier (**Subheading 3.5, steps 15–20**).

8. Purify all the FRET-peptides by HPLC and verify by MALDI-TOF. The calculated mass of the FRET-peptides should match the determined mass of the FRET-peptides.

9. Select the FRET-peptide with highest purity for validation and conjugation studies.

3.9. Antibody Conjugation

Like 2-iminothiolane conjugation approach, this method has two steps. First, the side-chain amino group of lysines of the antibody is derivatized with methyl-ketone group followed by conjugation of aminooxy derivatives of DOTA-ODC linker or biotinylated ODC linker (**Fig. 8**).

1. Equilibrate Sephadex-G100 with conjugation buffer, either triethanolamine 40 mM (pH 7.8) or phosphate buffer 50 mM (pH 7.8).

2. Fill 500 μL of already buffer-equilibrated Sephadex-G100 in minElute™ spin columns.

3. Purify required amount of antibody (10 mg/mL) by passing through molecular sieving column to remove impurities and equilibrate antibody with conjugation buffer.

4. Set up ketone linker (**Subheading 2.5**) and antibody conjugation at different molar ratios (10:1, 20:1, and 30:1) and rotate for 1 h at room temperature.

5. After conjugation, remove the excess free linkers by passing through the Sephadex-G100 spin columns.

6. In the second step, add aminoxy functional derivative at different molar ratios (10:1, 20:1, and 30:1) in TEA 40 mM (pH

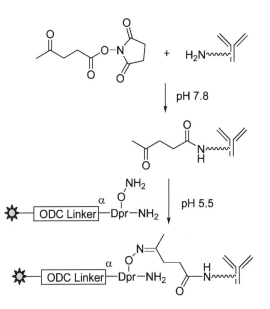

Fig. 8. Schematic of ketone–oxime conjugation method. Step 1: Conjugation ketone linkers to primary amines in monoclonal antibody. Step 2: Ketone–oxime bond formation between aminoxy functional group of ODC linker and ketone functional group of ketone linker.

5.5) to conjugate with ketone linker conjugated antibody and rotate for 4 h at room temperature.

7. We have selected a biotin-tagged protease tissue plasminogen activator (t-PA)-specific cleavable peptide as our functional derivative to conjugate to the antibody.

8. Remove the unconjugated functional derivatives by passing through Sephadex-G100 columns equilibrated with 10 mM disodium phosphate buffer, pH 8.0.

9. The amount of peptide conjugation per antibody can be determined by mass using MALDI-TOF mass spectrometry.

3.10. Protease Cleavage Studies (Notes 4.3 and 4.4)

1. Evaluation of protease digestion on peptide conjugate antibody can be done by either western blot analysis, immunohistochemistry, or flow cytometry before taking it to in vivo study. Following protocol is described for biotin-tagged peptide conjugated to an antibody.

2. Take 1 μg of antibody and biotin-peptide conjugated Ab with an appropriate enzyme buffer.

3. Incubate with protease (microunits to milliunits) for 1 h at 37°C.

4. Digestion can be evaluated by western blot analysis and IHC as described later.

3.11. Western-Blot Analysis

1. Make 10% SDS-PAGE gel as described in the procedure.

2. Load 1 μg/mL of ChL6 antibody, biotin-tagged peptide-conjugated ChL6 and enzyme-digested ChL6 antibody, and biotin-tagged peptide-conjugated ChL6 with protein markers.

3. Use biotinylated protein marker for standard ladder because it serves as a good control to check for protein transfer on membrane.

4. Separate on 10% SDS-PAGE gel at 100 V for 90 min and transfer to nitrocellulose membrane.

5. Block the membrane with 5% BSA for 1 h at RT. Alternatively, the blot can be blocked overnight with 1% BSA.

6. Incubate the blot with HRP conjugated avidin at a concentration recommended by manufacturer for 30 min.

7. Wash three times with TBS containing 0.2% Tween-20 to remove unbound avidin.

8. Develop the membrane by chemiluminescence reagent (ECL) according to manufacturer's instruction.

3.12. Immuno-histochemistry

1. Seed the cells in a six-well plate 1 day before the experiment. We have used ChL6, human chimeric antibody, to target breast adenocarcinoma cancer cell line HBT.

2. Keep the cells on ice for the duration of the experiment and use ice-cold reagents all the time. This helps the cells to adhere to plates.

3. Alternatively, the cells can be fixed with ice-cold methanol for 15 min and washed 3× with PBS. Block the cells by incubating with 1% BSA.

4. Block the cells by incubating with 1% BSA. This step can be skipped if the suspension cells are used for the study.

5. Add 1 μg/mL of ChL6 and peptide-conjugated ChL6 directly in 1 mL of growth media in duplicates and incubate on ice for 1 h.

6. Remove unbound antibodies by washing three times with PBS containing 1% BSA and 10 mM Mg^{2+}.

7. Add FITC-conjugated streptavidin (concentration as recommended by manufacturer) to each well and incubate on ice for 30 min.

8. Wash the plates 3× with ice-cold PBS containing 1% BSA to remove the unbound streptavidin.

9. Observe staining under the fluorescence microscope in green filter.

3.13. Flow Cytometry

1. Pass the cells 1 day before the experiment.

2. Distribute suspension cells (0.5 million cells per tube).

3. If the targeting cells are adherent cells, detach the cells by incubating with PBS containing 10 mM EDTA. Incubation time varies with cell type.

4. Wash the cells 3× with PBS to remove EDTA. This is very important because free divalent cations are necessary to target integrin receptor–ligand interaction.

5. Count the cells and aliquot them in Eppendorf tubes at 0.5 million cells per tube in 100-μL volume. Keep the cells on ice till the end of the experiment and use ice-cold reagents all the time.

6. Block the cells by incubating with 1% BSA.

7. Add 1 μg/mL of ChL6 and peptide-conjugated ChL6 directly in 1 mL of growth media in duplicates and incubate on ice for 1 h.

8. Remove unbound antibodies by washing three times with PBS containing 1% BSA and 10 mM Mg^{2+}.

9. Add FITC-conjugated strepavidin (concentration as recommended by the manufacturer) to each tube and incubate on ice for 30 min.

10. Wash 3× with ice-cold PBS containing 1% BSA to remove the unbound strepavidin.

11. Observe mean fluorescence intensity of the cell by running flow cytometry.

4. Notes

1. The experimental procedures described earlier can be used to identify protease-specific substrates. FRET-OBOC technology, by its nature, allows one to screen thousands to millions of substrates at a time (7, 29). It is important to consider enzyme mechanism of action and its biological characteristics while designing the libraries. For example, enzymes from serine proteases family required basic amino acids in the substrates and so on (**Subheading 1.3** and **3.1**).

2. Another advantage of bead combinatorial libraries is subtraction or removal of unstable compounds in biological fluids before adding the protease of interest (4). This process makes it possible for identification of specific substrates for proteases that belong to the same family or different isoforms of the same protease. We successfully used this technology to identify specific substrates for structurally and functionally similar serine proteases such as tissue plasminogen activator (t-PA) and urokinase (u-PA) (**Subheading 3.2** and **3.6**).

3. Protease cleavable substrates have shown promising results in clinical trails for delivering prodrugs to cancer cells especially, cathepsin and matrix metalloproteases *(30–32)*. These prodrugs are conjugated to a cancer cell-specific antibody. Internalized prodrugs are cleaved by cathepsin and liberate cytotoxic drugs to kill cancer cells. Research studies also showed that protease cleavable substrates can be used for tumor imaging by selecting substrates that are cleaved by tumor site-specific proteases such as matrix metalloproteases *(33–35)*.

4. Our proposed approach of using exogenous protease (e.g., TNKase) to cleave and therefore remove radiometal from cancer targeting antibody or agent will lower the unwanted radiation exposure to normal organs (**Subheading 3.8–3.10**). This brief protocol describes how highly specific protease substrates for these various applications can be identified.

Acknowledgements

The authors would like to thank Ekama Onofiok and Urvashi Bharadwaj for proofreading and funding support from NIH PO1CA047829-15A2.

References

1. DeNardo, G. L. (2005) Concepts in radioimmunotherapy and immunotherapy: Radioimmunotherapy from a Lym-1 perspective. *Semin. Oncol.* 32, S27–S35.

2. Vriesendorp, H. M., Quadri, S. M., Andersson, B. S., and Dicke, K. A. (1996) Hematologic side effects of radiolabeled immunoglobulin therapy. *Exp. Hematol.* 24, 1183–90.

3. Bennett, J. M., Kaminski, M. S., Leonard, J. P., Vose, J. M., Zelenetz, A. D., Knox, S. J., Horning, S., Press, O. W., Radford, J. A., Kroll, S. M., and Capizzi, R. L. (2005) Assessment of treatment-related myelodysplastic syndromes and acute myeloid leukemia in patients with non-Hodgkin lymphoma treated with tositumomab and iodine I131 tositumomab. *Blood* 105, 4576–82.

4. Kumaresan, P. R., Natarajan, A., Song, A., Wang, X., Liu, R., DeNardo, G., DeNardo, S. J., and Lam, K. S. (2007) Development of Tissue Plasminogen Activator Specific "On Demand Cleavable" (ODC) Linkers for Radioimmunotherapy by Screening One-Bead-One-Compound Combinatorial Peptide Libraries. *Bioconjugate chemistry ASAP Article* 10.1021/bc0602681 S1043-1802(06)00268-0.

5. Beeson, C., Butrynski, J. E., Hart, M. J., Nourigat, C., Matthews, D. C., Press, O. W., Senter, P. D., and Bernstein, I. D. (2003) Conditionally cleavable radioimmunoconjugates: A novel approach for the release of radioisotopes from radioimmunoconjugates. *Bioconjug. Chem.* 14, 927–33.

6. DeNardo, G. L., DeNardo, S. J., Peterson, J. J., Miers, L. A., Lam, K. S., Hartmann-Siantar, C., and Lamborn, K. R. (2003) Preclinical evaluation of cathepsin-degradable peptide linkers for radioimmunoconjugates. *Clin. Cancer Res.* 9, 3865S–3872S.

7. Kumaresan, P. R. and Lam, K. S. (2006) Screening chemical microarrays: Methods and applications. *Mol. Biosyst.* 2, 259–70.

8. Lam, K. S. (1998) Enzyme-linked colorimetric screening of a one-bead one-compound combinatorial library. *Methods Mol. Biol.* 87, 7–12.

9. Lam, K. S., Lehman, A. L., Song, A., Doan, N., Enstrom, A. M., Maxwell, J., and Liu, R. (2003) Synthesis and screening of "one-bead one-compound" combinatorial peptide libraries. *Methods Enzymol.* 369, 298–322.

10. Liu, R., Marik, J., and Lam, K. S. (2003) Design, synthesis, screening, and decoding

of encoded one-bead one-compound pepti-domimetic and small molecule combinatorial libraries. *Methods Enzymol.* 369, 271–87.

11. Chan, W. C. and White, P. D. *In: Fmoc Solid Phase Peptide Synthesis, Chan,* W. C. and White, P. D., eds. The Practical Approach Series; Oxford University Press, New York, pp 1–74.

12. King, D. S., Fields, C. G., and Fields, G. B. (1990) A cleavage method which minimizes side reactions following Fmoc solid phase peptide synthesis. *Int. J. Pept. Protein Res.* 36, 255–66.

13. Schmid, J. A. and Sitte, H. H. (2003) Fluorescence resonance energy transfer in the study of cancer pathways. *Curr. Opin. Oncol.* 15, 55–64.

14. Kohl, T., Heinze, K. G., Kuhlemann, R., Koltermann, A., and Schwille, P. (2002) A protease assay for two-photon crosscorrelation and FRET analysis based solely on fluorescent proteins. *Proc. Natl. Acad. Sci. USA* 99, 12161–6.

15. Martin, S. F., Hattersley, N., Samuel, I. D., Hay, R. T., and Tatham, M. H. (2007) A fluorescence-resonance-energy-transfer-based protease activity assay and its use to monitor paralog-specific small ubiquitin-like modifier processing. *Anal. Biochem.* 363, 83–90.

16. Lam, K. S., Lebl, M., and Krchnak, V. (1997) The "one-bead-one-compound" combinatorial library method. *Chem. Rev.* 97, 411–48.

17. Lam, K. S., Salmon, S. E., Hersh, E. M., Hruby, V. J., Kazmierski, W. M., and Knapp, R. J. (1991) A new type of synthetic peptide library for identifying ligand-binding activity. *Nature* 354, 82–4.

18. Meldal, M., Svendsen, I., Breddam, K., and Auzanneau, F. I. (1994) Portion-mixing peptide libraries of quenched fluorogenic substrates for complete subsite mapping of endoprotease specificity. *Proc. Natl. Acad. Sci. USA* 91, 3314–18.

19. Cannon, C. P., McCabe, C. H., Gibson, C. M., Ghali, M., Sequeira, R. F., McKendall, G. R., Breed, J., Modi, N. B., Fox, N. L., Tracy, R. P., Love, T. W., and Braunwald, E. (1997) TNK-tissue plasminogen activator in acute myocardial infarction. Results of the thrombolysis in myocardial infarction (TIMI) 10A dose-ranging trial. *Circulation* 95, 351–6.

20. Ding, L., Coombs, G. S., Strandberg, L., Navre, M., Corey, D. R., and Madison, E. L. (1995) Origins of the specificity of tissue-type plasminogen-activator. *Proc. Natl. Acad. Sci. USA* 92, 7627–31.

21. Guerra, D. R., Karha, J., and Gibson, C. M. (2003) Safety and efficacy of tenecteplase in acute myocardial infarction. *Expert Opin. Pharmacother.* 4, 791–8.

22. Kaiser, E., Colescott, R. L., Bossinger, C. D., and Cook, P. I. (1970) Color test for detection of free terminal amino groups in the solid-phase synthesis of peptides. *Anal. Biochem.* 34, 595–8.

23. Vojkovsky, T. (1995) Detection of secondary amines on solid phase. *Pept. Res.* 8, 236–7.

24. Song, A. M., Wang, X. B., Zhang, J. H., Marik, J., Lebrilla, C. B., and Lam, K. S. (2004) Synthesis of hydrophilic and flexible linkers for peptide derivatization in solid phase. *Bioorg. Med. Chem. Lett.* 14, 161–5.

25. Liu, R. W. and Lam, K. S. (2001) Automatic Edman microsequencing of peptides containing multiple unnatural amino acids. *Anal. Biochem.* 295, 9–16.

26. King, D. S., Fields, C. G., and Fields, G. B. (1990) A cleavage method which minimizes side reactions following Fmoc solid-phase peptide-synthesis. *Int. J. Pept. Protein Res.* 36, 255–66.

27. Greg, H. T. (1995) *Bioconjugate Techniques,* Academic, San Diego.

28. Moses, J. E. and Moorhouse, A. D. (2007) The growing applications of click chemistry. *Chem. Soc. Rev.* 36, 1249–62.

29. Meldal, M. (2002) The one-bead two-compound assay for solid phase screening of combinatorial libraries. *Biopolymers* 66, 93–100.

30. Warnecke, A., Fichtner, I., Sass, G., and Kratz, F. (2007) Synthesis, cleavage profile, and antitumor efficacy of an albumin-binding prodrug of methotrexate that is cleaved by plasmin and cathepsin B. *Arch. Pharm.* 340, 389–95.

31. Potrich, C., Tomazzolli, R., Dalla Serra, M., Anderluh, G., Malovrh, P., Macek, P., Menestrina, G., and Tejuca, M. (2005) Cytotoxic activity of a tumor protease-activated pore-forming toxin. *Bioconjug. Chem.* 16, 369–76.

32. Harada, M., Sakakibara, H., Yano, T., Suzuki, T., and Okuno, S. (2000) Determinants for the drug release from T-0128, camptothecin analogue-carboxymethyl dextran conjugate. *J. Control Release* 69, 399–412.

33. Deguchi, J. O., Aikawa, M., Tung, C. H., Aikawa, E., Kim, D. E., Ntziachristos, V., Weissleder, R., and Libby, P. (2006) Inflammation in atherosclerosis: Visualizing matrix metalloproteinase action in macrophages in vivo. *Circulation* 114, 55–62.

34. Melancon, M. P., Wang, W., Wang, Y., Shao, R., Ji, X., Gelovani, J. G., and Li, C. (2007) A novel method for imaging in vivo degradation of poly(l-glutamic acid), a biodegradable drug carrier. *Pharm. Res.* 24, 1217–24.

35. Jiang, T., Olson, E. S., Nguyen, Q. T., Roy, M., Jennings, P. A., and Tsien, R. Y. (2004) Tumor imaging by means of proteolytic activation of cell-penetrating peptides. *Proc. Natl. Acad. Sci. USA* 101, 17867–72.

Printed in the United States of America